PARIS JOURNAL

1944-1955

PARIS JOURNAL

VOLUME ONE
1944-1955

❖

JANET FLANNER
(GENÊT)

❖

EDITED BY WILLIAM SHAWN

A Harvest Book

HARCOURT BRACE & COMPANY

San Diego   New York   London

Copyright © 1965, 1955 by The New Yorker Magazine, Inc.
Copyright 1954, 1953, 1952, 1951, 1950, 1949, 1948, 1947,
1946, 1945, 1944 by The New Yorker Magazine, Inc.

Paris Journal, Volume 1 1944–1955
and *Paris Journal, Volume 2 1956–1964*
were previously published together as
Paris Journal, Volume 1 1944–1965.

Library of Congress Cataloging-in-Publication Data
Flanner, Janet, 1892–1978
Paris journal.
"A Harvest book."
Includes indexes.
Contents: v. 1. 1944–1955 — v. 2 1956–1964 —
v. 3. 1965–1970.
1. Paris (France)—History—1944–   . 2. France—
Politics and government—1945–   . 3. Paris (France)—
Intellectual life—20th century. I. Shawn, William.
II. Title.
DC737.F55 1988   944'.361082   88-2300
ISBN 0-15-670948-1 (v. 1)
ISBN 0-15-670949-X (v. 2)
ISBN 0-15-670952-X (v. 3)

Printed in the United States of America

First edition

B C D E F

To my editor, William Shawn,

with admiration, gratitude,

and more than affection

PARIS JOURNAL

1944-1955

1944

December 15

This is a period of high visibility in France. The surface of Paris today consists of fundamentals, as it has before during the country's crucial mutations. The good and bad that lie at the base of French life are now being temporarily pressed to the top. Today is a general French date, like the date after the French Revolution, like the date after the fall of Napoleon. On such a calendar, it seems evident, 1918 was no date at all.

What, each according to his station in life, the citizens of Paris, four months after its liberation, hope that they will live to see in the way of a revived and reformed France, and what, rich or poor, they know they cannot live through again are doubtless clearer now than they will be later, when normal living and forgetfulness begin to dim the scene. Existence in Paris is still abnormal with relief, with belief. The two together make for confusion. The population of Paris is still a mass of uncoördinated individuals, each walking through the ceaseless winter rains with his memories. Government, because it is a novelty after four years of occupation, seems so intimate that each citizen feels he can keep his eye on it in vigilant curiosity. News, too, is intimate, as if the globe had shrunk to suit the size of the one-sheet French newspapers. Large events, when reduced to a paragraph, seem personal, and battlefronts on which millions of men of various nationalities are struggling assume, in print, the intense, direct interest of a family affair in which three of its members are involved, since France metaphysically considers that she has three armies—her soldiers, her deportees, and her tortured civilian dead. No Parisian, worker or capitalist, patriot or turncoat, important or obscure, has

3

enough of everything. Nourished by liberation, warmed by the country's return to active battle, Paris is still, physically, living largely on vegetables and mostly without heat. France, which was the strained and straining leader of Europe for so many years that finally she no longer had the strength to lead herself, is now probably the most fortunate survivor of an Old World that is being destroyed into something new.

For five years Europe has been the victim of cannibalism, with one country trying to eat the other countries, trying to eat the grain, the meat, the oil, the steel, the liberties, the governments, and the men of all the others. The half-consumed corpses of ideologies and of the civilians who believed in them have rotted the soil of Europe, and in this day of the most luxurious war machinery the world has ever seen, the inhabitants of the Continent's capital cities have been reduced to the primitive problems of survival, of finding something to eat, of hatred, of revenge, of fawning, of being for or against themselves or someone else, and of hiding, like savages with ration cards. The desperate economic competition which will arrive with the peace will be scarcely less bloodthirsty. The country whose men, machines, and municipalities are the least wrecked physically; the country lucky enough to have unweary, educated brains, young or old, which will volunteer their services to public life, like youthful corporals or middle-aged colonels volunteering for a dangerous, unrewarded front-line mission; the country which, above all, has in readiness a leader not identified with military defeat or, even worse, what peacefully led up to it—that country should have the best chance for survival. At this moment of semi-liberation in Europe, with the end of hostilities surely situated somewhere across the Rhine, in a field or village not yet in history's view, France seems the Continental nation most favored for the future. As *Combat,* a former underground newspaper that has assumed great importance here, has just declared, this is the time, in France, for harsh intelligence. As *Figaro,* which represents a new, noble-hearted Right, has just stated, the French must rethink many things which up to now they have failed to think to a conclusion. As the Communist *Humanité* repeats in its slogan, which sounds strangely like an imperial throwback, it is now or never for the "French Renaissance."

Until the Renaissance gets under way, mouths and hands are more important than intellects and hopes. The shortages of food and employment are equally great; there is a nice balance between the empty factories and the nearly empty stomachs. When the Germans disappeared from Paris in August, the black market began to disappear, too. This has not turned out to be the blessing one might have expected. The black market, and even the gray market—the constant bartering, selling, and haggling that went on among friends—devastated what was left of French morale and cheapened the value of time, since it consumed hours of each day, but it also saved French lives. The black market was built upon the French peasant's classic cupidity and the Vichy government's stupidity in thinking that it is a farmer's social duty to grow things whether they profit him or not. Because the French peasant was paid parsimoniously low official prices in an attempt to keep consumer prices down, he found that he was receiving only a fraction of what it cost him to produce what he sold. Therefore, humanly, he not only lied to the authorities about the quantity he could bring to market but, inhumanely, he undersupplied even his assessed regional share for the needy. And while the Parisian, in the four years of the occupation, was losing, on the average, forty pounds in weight because of inadequate rations, the peasant was selling his surplus on the black market at profits which, if he had big herds, flocks, and fields, made him a franc millionaire. Now, as one of the money-minded *nouveaux riches,* he fears the inflation he helped to create. Thus nowadays a cow in the bush seems a better bet than thirty thousand francs in the hand. The peasant has stopped selling, the black market is drying up, and Parisians, who expected to fill up on something more than freedom after the liberation, are still underfed. This week's fat-and-flesh ration for a family of three in Paris is a half pound of fresh meat, three-fifths of a pound of butter, and nine twenty-fifths of a pound of sausage. On this much food the family of three will not starve, but neither will any member of it feel full. Since the breakup of the excellently organized black market, substitute arrangements have come into being and with them have come hijacking, robberies, connivances, and Chicagoesque gangs drawn from the Mauvais Maquis, which was composed of thugs who had nothing to do with the original Maquis (the Communist Maquis Rouge) or the subsequent non-political Maquis Blanc and who have done the properly sacred Maquis legend harm. It is nothing now for helpless officials to announce that en route from Normandy to Paris two tons of butter have melted from a six-ton truckload. For that matter, whole trainloads of food have been reported lost, including the locomotive.

The first of the two major attacks by the Consultative Assembly on de Gaulle's Cabinet to date was made by slightly hungry Deputies, who asked questions about supplies. Last year, under the Nazis, the Deputies pointed out, there were turkeys for those who could pay a thousand francs for them. This year there were only bombed bridges, too few freight cars, and blasted railroad tracks. The Germans departed on and with two-thirds of the country's railway equipment and two-thirds of the country's trucks. This year, too, on the war maps, there is a great food-producing slice of France which even the Deputies may have forgotten about, since it is never mentioned and is still controlled by the Germans—from the lamb-grazing pasture lands above Bordeaux up into southern Brittany, secondary butter-and-egg basket of the country. Anyway, even if there were quantities of food in Paris, there would be nearly no fuel with which to cook it. The coal is in the north. There can be no transportation without coal and no coal without transportation. Two months of continuous rain have so swollen the rivers and canals that coal barges are tied up to the banks. Because of the lack of coal, there is no electric power for the city's factories, so they remain closed and the men have no work to do. But if they were open, who would fill all the jobs? Eight hundred thousand of the most skilled French factory workers are still labor slaves in Germany, working against the Allies. Because of the lack of electricity, a dozen of the Métro stations have been closed in an effort to force people to walk and reduce the number of passengers. The government has just announced that no public buildings will be heated before January 1st. The Academicians, old and wise men, have just announced that they will not hold their end-of-the-year annual meeting until some nice day next spring.

Then, because of the coal shortage, there is a gas shortage. Gas for cooking is rationed to an hour and a half for luncheon—it is on between noon and one-thirty—and an hour for supper, between seven and eight. There is no gas to heat the breakfast coffee of burned barley. Parisians rich enough to buy sawdust by the ton, and black-market gasoline for trucks to deliver it, can store a winter's supply of several tons of sawdust in half of an apartment and keep the other half warm with special sawdust-burning stoves. With equally special low-current electrical attachments, the newly invented stone-lined stoves here can be heated during the night, after which the hot stones give warmth most of the next day. But these attachments are no longer available. For nearly everybody in Paris, the only available commodity is the cold. Except for the first winter after the defeat, this is the most uncomfortable winter of the war. Nevertheless, to the freed Parisians, dining in their overcoats on a meagre soup of carrots and turnips, the only obtainable vegetables, Paris seems like home for the first time since the Christmas before the war.

The second attack by the Consultative Assembly on the de Gaulle Cabinet was on the question of *Epuration,* as the cleaning up of the collaborationists is called. The appetite of many people in France for justice, too, is unsatisfied. Certainly there are families to whom the word "justice" will lose all meaning, except to serve as a mocking epitaph for the tombs of sons, husbands, and even daughters and wives—provided their bodies could be found to be buried—unless some traitor's life is taken (legally and quickly, by a firing squad) in payment for each patriot's life (taken slowly, behind locked doors and in agony). During the early, idealistic, inexperienced days of the resistance, a member of a group was trusted never to betray his comrades, who therefore did not disband and disperse if he was caught. Statistics later proved that the average member could not resist the Nazi tortures more than twelve hours, which was all, finally, that his comrades asked of him, the half day being sufficient for them to hide. *Le Franc-Tireur,* the resistance paper that has remained most faithful to the memory of the dead, has just candidly demanded a thousand heads and five thousand arrests to wipe out the old score. On the other hand, eloquent, calm pleas against repeating what in France's Revolutionary days was called *Le Terreur* have been made, usually by good citizens not in mourning. There sometimes seem to be few who are not. In two families your correspondent knew well, an only son, in each instance a Sorbonne student, was killed in the street fighting on the Place St.-Michel during the liberation of Paris. A fashion editor's daughter, a delicate, blond figure in the early intellectual resistance, lived through a year's solitary confinement in a cell in the Santé Prison and today is a slave factory worker in Germany. A French Catholic poet, great-nephew of a cardinal and betrothed when the war broke out to an American Jewish girl, officially declared himself an *Enjuivé,* or Semitic sympathizer, then lived in hiding for four years without ration cards. He is now an eighty-five-pound skeleton. The son of an anti-Nazi refugee German woman intellectual, who aided the son to flee to England and the R.A.F., became a parachutist and made jumps in France to obtain information for the British about the resistance. Just before the Nazis left Paris, they captured him in the suburbs, and before taking him off, they tortured him, mutilating his hands. Behind the cushions of a French police car into which the Germans put him after the torture, he left a blood-stained note addressed to his mother. She does not know who mailed it to her. One of the best-known women journalists of Paris, who refused to be party to the collaboration of her old paper, *Le Matin,* is crippled by rheumatic fever from having lived for years, without a job, in an unheated house. A Jewish singing teacher developed such severe chilblains on her hands in the Jewish concentration camp at Drancy that they turned black. They are still too swollen for her to play scales for her pupils. A White Russian, sister of a famous painter, became a seamstress, a Communist, and a sufferer from tuberculosis in the course of the war. A composer of modern music who had been aiding the resistance was deported, after being arrested by the Gestapo at a house where he had called by mistake and where the Germans were seeking a man he did not know and who had already fled. In the Rue Jacob, where a printer was seized, to be shot, for setting type for an underground resistance paper, Gestapo agents discovered a wedding list left in the shop by a matron who had ordered invitations for her daughter's marriage; all the people on the list were arrested. One American woman, in her château near Chartres, had to quarter the officers of the local German command in her downstairs rooms for two months while she hid five men of the Maquis on an upper floor; the two groups never met, but her hair began to turn white. One music-loving New Yorker with a small country cottage not far from Paris hid and fed, by her own labors on her patch of land, one singer and one church organist wanted by the Gestapo for helping the Maquis, one French poet wanted by the Paris police, and one Polish deserter from a German artillery regiment camped on a farm above her little valley. In the past week, three friends from the old days have written your correspondent letters on black-edged stationery but have not said what their loss is.

French politics, unfortunately, will have to start up again soon. Local elections for the Municipal and General Councils will be held in February, the first balloting since the liberation and the first in the country's history in which women will cast their vote. To make it easier for them, they are being registered as they pick up their December food-ration cards. They will outvote the males of France by four million. Two and a half million Frenchmen are, of course, disfranchized, being still sealed inside Germany. So far, General de Gaulle has no political party; instead, he has popularity. What program he has is social; his slogan is *"Rénovation,"* which means reforming France out of the nineteenth century into the twentieth. Even people who are against him believe that he has been the savior of France. The grand old political parties have not yet raised their heads or voices to say what they think of *rénovation*. Nobody in France any longer has the courage to describe himself as a conservative and at the other end of the line there is also lacking a party of revolution. The Communists are now, realistically, the party for construction. The Communists, who were the great heroes of the resistance, are also, as the result of their excellently organized underground years, the best-organized party in France. They have just accepted affiliation with the weakened Socialist group, as a healthy man might take on an invalid wife. French Communism seems to be something very different from what it was in the old international days, since now it is avowedly pro-French. On the extreme right is what remains of the rich owner class, who seem readier than they were during the Front Populaire for a sort of belated New Deal. This they describe as Saint-Simonisme, which calls for a Christian, responsible élite, with precepts laid down by the French social philosopher of more than a century ago, Claude-Henri Saint-Simon.

There has been a great national deviation from morality as the result of the French people's contact with the Nazi doctrines and of the French people's unhappy, defeated, introverted contact with itself. The materialism of the nineteenth century, the *enrichissez-vous* of the Bourbon revival, the formal corruption of the kings named Louis all established habits which the Third Republic inherited and which now only silent anti-democratic French imbeciles can suppose began with the Third Republic's first ballot. Citizens who vote have to have political parties to vote for, but what France needs most, now that she clearly has a leader without a past, is not the salvation offered by any one political party but a revival of morality, to be practiced alike by the governing and the governed.

1945

January 1

The most appropriate decoration on the rare Parisian Christmas tree this year was ribboned tinfoil, which American bombers cast into the suburban air when towns like Le Pecq, Achères, and Poissy were being softened up for the liberation of Paris. These fragments, intended merely to interfere with German radio-detection devices, were salvaged by trinket-hungry French children and ended up as décor for the wartime holiday whose slogan was still "Peace on earth."

Charity drives are now under way to help the many thousands of French who have lost only their homes and not their lives and whose dwellings are now debris-filled cellars in Le Havre, Dieppe, Caen, Rouen, and other towns where, without the added melodrama of enemy sadism, the victorious push of the Allies was enough to bring ruin. The drives are for food, but there is little for anyone to give, and for clean old clothes, of which there seems to be none. A motion has been made in the Consultative Assembly to set up a Ministry of Victims of the War. A Ministry of Prisoners, Deportees, and Refugees already exists. Under its sponsorship was celebrated, in the days between Christmas and New Year's, *La Semaine de l'Absent,* in honor of that twentieth of France's population which is still prisoner, one way or another, in Germany. The posters for the absent ones bore nothing but de Gaulle's Lorraine Cross, made of twisted barbed wire. These are the images, these are the texts, these are the greetings from life which took the place of 1944 Christmas cards in Paris.

January 17

Liberated Paris is now occupied by snow. Parisians are colder than they have been any other winter of the war. They are hungrier than they have been any other winter of the war. They are the hungriest they have been since the Prussian siege of Paris, when their grandparents ate mice. Because there is still a lack of coal to transport anything, even coal, the citizenry is still waiting, muffled to its ears, for its single sack per person due last August. Electricity has just been ordered turned off all over France from morning till evening, except for an hour at noon, perhaps for the women in dark kitchens to see what little they are cooking. For fear that the plumbing will burst, the water is turned off in many Paris apartment houses and office buildings except between noon and two o'clock, too late to do most people any good. Passenger trains pulled by steam engines have been suppressed. Only trains pulled by Diesel engines or electricity will now operate. People who must travel a distance of more than a hundred kilometres must get a permit, which will probably take them more time than if they walked.

Owing to the paper shortage, announced by a paper trust which was able to find enough for the Nazis during the occupation, Parisian newspapers, which were already limited to one sheet, are now down to a half-size sheet. There is even a shortage of salt. Lacking salt, whatever Normandy butter could be stored is turning rancid. Lacking salt, Breton peasants don't kill their pigs, which makes for a lack of *charcuterie,* the secondary meat ration in France. Lacking everything but rain in the autumn, Ile-de-France farmers were unable to pull their sugar beets out of the mud, so that there is a shortage of sugar and there may soon be a shortage of cows, because to feed a French cow on beet root now costs a hundred francs a day, which used to be the price of dinner at the Ritz. Because of the beet-root shortage, many a French cow will not live to see the pleasant spring or her pretty calf. As one butcher gloomily declared, it will be not slaughter but a double assassination. Owing to the grain shortage and the fact that under the Germans the French were eating in their bread some of what their horses usually ate in the manger, many an old French horse will not be allowed to last out the winter, either. However, owing to the gasoline shortage, even last month's ration of horse meat has not yet been brought to town.

Milk is now available only to what the ration books classify as "J-1," which means *"Jeunesse Première,"* or babies of from one to three, who are also in line this month for their one annual dried banana, if it arrives. January also entitles "J-2" (children of from four to fourteen) and "J-3" (adolescents up to twenty-one) to four eggs apiece, which their parents, if they can afford it, will probably buy on the black market at thirty francs each. Clinics, filled with pneumonia victims from otherwise charming homes, lack drugs and heat. The nurses lack thermometers, the surgeons lack light to operate, the dentists, lacking electricity, for once cannot drill. There is no plaster-of-Paris in the hospitals to set the record number of bones that are being broken because of falls on icy streets. Otherwise, French doctors say, people's bones are almost too soft to break, because of decalcification due to undernourishment in general and especially to the lack of fish, which the Nazis forbade the Bretons to fish for in the pre-invasion Channel waters. The invasion is over, but there is still no fish in Paris. There is nothing of lots of things which could be called normal, even in war.

Little is left even of some of the more orderly high hopes which fluttered in French hearts as the French flags fluttered overhead on the day of liberation. Paris is free, and no one forgets it for a minute, since nothing much else has happened to remember. Because the person of General de Gaulle is still sacrosanct, the editorials do not criticize him, but every day they carp at his provisional government as if he had nothing to do with it, which is beginning to seem to be the case. His Assembly is one of the strangest on earth today, since it contains, among others, representatives of four political groups named after four Left Wing newspapers. Each of these groups has six delegates and represents the editorial faith of one of the four powerful anti-Pétain, pro-patriot underground papers of the occupation—*La Libération du Nord, La Libération du Sud, Le Franc-Tireur,* and *Combat.* It is as if New York had a couple of Leftist congressmen whose party was called *PM,* or as if the isolationist senator from Illinois had had to drop the Republican tag and come right out and say that he belonged to the Chicago *Tribune.* The Assembly also contains a hundred and five other patriot delegates who could but do not unify themselves into a majority—two groups called Those of the Resistance and Those of the Liberation (the leading ex-underground groups, now organized to the extent of even having offices and telephone numbers), as well as the Socialist resistance group of twenty-one delegates, now, as before the war, the largest single party; the Communist Resisters; and, for the first time, a labor-union group, the Confédération Générale de Travail, which is the French C.I.O. Alone of the old domineering, inefficient, middle-of-the-road pre-Munich parties, the Radical Socialists have squeezed themselves into the picture by officially declaring that their prewar party policy was "one of dastardliness and desertion." The drollest fact about the de Gaulle political scene is that it is the arch-conservatives like Louis Marin, of the old Fédération Républicaine, who have now gone underground, like yesterday's revolutionaries.

Newspapers, no matter what their size, have never before been so highly regarded in France as they are today. In clandestinity, news was dangerous and rare and precious, and Parisians still don't have enough of it. The most remarkable new journalistic figure in France is Albert Camus, an editorial writer on *Combat*. He was born in Belcourt, a modest suburb of Algiers, thirty-one years ago. He majored in philosophy at the University of Algiers and became a neo-Sophist whose wisdom consisted of thinking that life was ridiculous. In this mood he wrote some not very wise poems and an interesting novel called "L'Etranger," which revealed a Hemingway influence in style but not in substance, since the point of the book was that the hero did not know anything about himself. Camus wrote a play, "Le Malentendu," which was produced in Paris under the Germans last spring and featured a man who so misunderstands himself that he commits a meaningless murder. Everything Camus writes today in *Combat* is incised with meaning. An ill young man (he is tubercular), he watches the available fragments of French news with the extra attention of the invalid, too intelligent and objective to presume that he will ever get the full story.

Of all the revolutions in the professions, whose reforms the resistance had hoped would become part of a changed, cleansed republic after the liberation, the journalists' revolution is the only one which seems to have succeeded, perhaps because, being active, intrepid, and young, they established their setup before anyone, in the early excitement of freedom, had time to think twice and stop them. They gained their two major objectives—the right, as journalist patriots without printing plants, to take over the printing plants of the nonpatriot Paris press, whose owners had fled when the liberation came; and the right, as a new, thinking political nucleus, to become delegates to the Assembly in that new, free France for which they had resisted and bled. Since freedom arrived, they are the only examples here of someone's having reached for the moon and got it. So the new Republic has started off with something the old did not have—the most intelligent, courageous, and, *grâce à Dieu,* amateur press which venal, literate France has ever known. The new press of Paris alone lives up to the resistance slogan, *"Les Durs."* It is indeed hard, and pure.

A new crime has just been created, by ordinance. It is called *l'indignité nationale* and it certainly covers a multitude of collaborationist sins. Roughly, to have committed national indignity, a Frenchman need only have belonged to "that government in authority"—and recognized by Washington—between June, 1940, and the provisional government of today; or have sung songs, spoken poetry, or even helped organize the former government's many artistic galas. The punishment for anyone found guilty of national indignity is equally comprehensive and is called *dégradation nationale.* National degradation will consist of being deprived of nearly everything the French consider nice—such as the right to wear decorations; the right to be a lawyer, notary, public-school teacher, judge, or even a witness; the right to run a publishing, radio, or motion-picture company; and, above all, the right to be a director in an insurance company or bank. As time goes on, fewer collaborationists are going to be shot and more of them, in their national degradation, will be looking for jobs.

February 21

The brightest news here is the infinite resilience of the French as human beings. Parisians are politer and more patient in their troubles than they were in their prosperity. Though they have no soap that lathers, both men and women smell civilized when you encounter them in the Métro, which everybody rides in, there being no buses or taxis. Everything here is a substitute for something else. The women who are not neat, thin, and frayed look neat, thin, and chic clattering along in their platform shoes of wood—substitute for shoe leather—which sound like horses' hoofs. Their broad-shouldered, slightly shabby coats of sheepskin—substitute for wool cloth, which the Nazis preferred for themselves—were bought on the black market three winters ago. The Paris midinettes, for whom, because of their changeless gaiety, there is really no substitute on earth, and who are what our seventy-two-hour-leave soldiers are out to catch or be caught by, still wear their home-made, fantastically high, upholstered Charles X turbans. Men's trousers are shabby, since they are not something which can be run up at home. The young intellectuals of both sexes go about in ski clothes. This is what the resistance wore when it was fighting and freezing outdoors in the *maquis,* and it has set the Sorbonne undergraduate style. Nothing is left of the proletariat male and female Zazous (opposite numbers to our zoot-suiters), who had their own special resistance sartorial getup—nothing, that is, except the long, crook-handled umbrella which all French women now carry, rain or shine. That was only a part of the strange chic of the female Zazous, along with no hat, an artful, absolutely terrific pompadour, a strict, tailored suit with a full skirt for bicycling, and, hanging from the neck, the inevitable cross. It was not a de Gaulle Lorraine cross; it was a Christian cross. Second-hand shops are now full of them and nobody can explain what they signified. Male Zazous were like the long-coated zoot-suiters except for their pantaloons, which were skintight; no one can explain that either.

The more serious normalities of traditional Paris life go on, in readjusted form. Candy shops display invitations to come in and register for your sugar almonds, the conventional sweet for French baptisms, but you must have a doctor's certificate swearing that you and your wife are really expecting. Giddy young wedding parties that can afford the price pack off to the wedding luncheon two by two in *vélo-taxis,* bicycle-barouches which are hired for hundreds of francs an hour. The other evening your correspondent saw a more modest bridal couple starting off on their life journey together in the Métro. They stood apart from everyone else on the Odéon platform, the groom in his rented *smoking* and with a boutonnière, the bride all in white—that is, a white raincoat, white rubber boots, white sweater and skirt, white turban, and a large, old-fashioned white nosegay. They were holding hands. American soldiers across the tracks shouted good wishes to them.

The year's first French literary prize has just been awarded and been tossed back into the donors' ladylike faces. It is the Prix Fémina-

Vie Heureuse, or the Fémina Happy Life Prize, unquestionably, despite its faded title, second in importance only to the Prix Goncourt, and it was given to the writers of twenty-one clandestine booklets published by the Editions de Minuit during the occupation. Good writing was one thing the intellectual resistance did succeed at. However, the midnight-edition writers, or such of them as are available, have not only scorned the happy-life prize but have insulted the jurors to boot. In a round-robin letter they accuse the prize-givers of never having "busied themselves to combat the invader or his accomplices. It is derisory for them to give an award of this sort to writers who were and are still fixed on the sublime goal of the Rights of Man and of whom some—let us not forget them—were killed or deported by the Germans." The letter was signed by, among others, Elsa Triolet, François Mauriac, Claude Aveline, Jean Guehenno, Louis Aragon, Jean Cassou, Claude Morgan, Paul Eluard, and Vercors. One major character who did not sign was M. Alaud, the little Place d'Italie printer who risked his life to print their works. His regular commercial specialty was printing black-edged funeral notices, and he was never suspected of also printing for the lyric heroes of death. The midnight-edition group is genealogically the descendant of Dada and then of Surrealism.

In the resistance's best writing, patriotism and talent together attained heights which are an important part of the new literary horizon of France. All the members wrote fiercely, with a finesse sharpened by danger and despair, all celebrated what they called *"l'époque des prisons,"* and all signed false names. The two who burned brightest in those dark years were Aragon, who signed himself François la Colère, and Paul Eluard, who wrote under many names or none at all. By the exciting extensions of his imagination, in "Le Musée Grévin," a poetical chamber of horrors in which the waxen figure of the Vichy Marshal (*"cette vieille terreur aux traits de macaron"*) appears, and in "Les Yeux d'Elsa," a group of poems addressed to his wife, Elsa Triolet, Aragon broke new ground for his rhymed anger and tenderness. Eluard's most celebrated creation is his poem "Liberté," written in 1942, which was secretly circulated through imprisoned Europe as a clandestine classic. As a dolorous feat, the "Trente-trois Sonnets Composés au Secret," by Jean Cassou, who called himself Jean Noir, remains unique. He composed them while in prison, where, since writing materials were forbidden, he was forced to incise them in his memory. Cassou was an unremarkable literary figure before the war; prison unlocked his talent. Under the nom de plume of Forez, even Academician François Mauriac contributed some stirring lines on the state of the French nation's soul. Elsa Triolet's popular novel "Le Cheval Blanc," one of the few works not on the Germans' Otto list, or proscription index—nobody knows who Otto was—established her as the leading new talented writer among French women.

Malraux avoided Otto by publishing his last volume, "La Lutte avec l'Ange," in Switzerland. The most popular freak piece of fiction and the most contentious, since its hero is a Siegfriedian German of the type the French saw few of, still remains "Le Silence de la Mer," a hymn signed by Vercors, the resistance name of Jean Bruller, publisher of the Midnight Editions. Before the war he published *cahiers* of his own unimportant drawings and he is now preparing another one. He wishes to be known as an artist rather than a writer, although he is a better writer. He was such an excellent resistance publisher that his twenty-one booklets, clandestine though they had to be, were elegantly printed on vellum paper, in numbered, limited, subscribed editions. Each booklet, which originally sold for twenty-one hundred francs, is now worth up to fifty thousand to collectors. A faithful reprint edition of all twenty-one booklets, on the same paper and at the original price, is now being published. Each volume bears, as a headpiece, the phrase "Printed by Alaud, clandestine printer." Each ends with a tailpiece which says, like a memo to history, "This volume achieved publication under the Nazi oppressor."

March 7

Paris, for the first time since the liberation, seemed herself again the morning after de Gaulle's recent speech to his Consultative Assembly. For the first time the Paris newspapers were in full critical cry. The brilliant clamor of the dissenting editorial opinions was a treat. It was like listening to the intelligent, uproarious voices of dear old friends and dear old enemies. To enable the journals to give full attention to the General's long-awaited speech, they were allowed, also for the first time, an extra ration of paper, so that they could print four pages instead of the usual two. The august mutism of the General had up to then constrained the French press and people to a polite taciturnity of their own; being mere citizenry, they were waiting for the oracle to speak up first. If the volubility which has followed the speech continues, Paris will seem gustier, more nearly normal, and closer to reality than it was. After any significant parliamentary debate in the days before the war, the friends and enemies in the press were always against each other and for somebody. In their reaction to the de Gaulle speech, the editors were still not for each other, but they were unanimous in being against de Gaulle. They were at odds only in their reasons.

The tipoff to the General as to what at least two papers expected of him was given the morning before his speech by the Communist *Humanité* and the Socialist *Populaire* in a jointly written three-quarter page *mot d'ordre* which demanded, in large type, the old, familiar resistance program—immediate nationalization of big banks, later nationalization of natural resources, key industries, and transport and insurance companies, and the confiscation of all traitors' earthly goods, "not as vengeance but as a moral exigence." Inevitably, neither was satisfied with what the General offered, though it sounded revolutionary. *L'Humanité* sniffed at national-izations in which the de Gaullist state would hold fifty-one per cent of a company's stock but the old directors could continue to own a forty-nine-per-cent interest. As for traitors' fortunes, *L'Humanité* declared itself "profoundly ulcerated" to see that the two hundred families of the Banque de France appeared to be above the law. It also said that if state ownership was Communism, then Louis XI was a Communist, because he founded the French post office. The Social-ist *Populaire* declaimed against the collaborationist bourgeoisie which had saved its investments but lost its honor.

Combat, perhaps the most intelligent of the resistance papers in its criticism of the General, said it had been "filled with fears and reservations as to what he might say, though by his words he encouraged our dearest hopes. When he ends words and starts acts, he will obtain the gratitude of the entire country." De Gaulle's speech was received coldly and almost without applause by all the resistance, trades-union, and Leftist benches.

If this had been the General's opening speech to his nation, it would have been historic. In its elegant, démodé vocabulary, which stiffened his ideas into classic stateliness, his address would have been perfectly suited to the Académie. The grand type of French he spoke and thought in put him two hundred years away from his listeners on the radio. In his formal phraseology, he impatiently deplored the lack of good brains for big tasks. He quoted Shakespeare. He spoke of the "twelve million handsome babies" which France must have in the next ten years, "for if the French people do not multiply, France will be nothing but a great light which is extinguished." He spoke of never having lulled France with illusions, spoke of the wisdom of seeing large and looking high, of his attempt to measure his own words, and of his desire that others should measure theirs. It was a splendid speech—to read afterward. When spoken in the Assembly, it pleased neither reformers nor conservatives, both of whom wish, for very different reasons, that he would stop talking and start doing, so that at least one side can know where it stands with him.

Never, perhaps, has the French journalist written with a greater sense of responsibility than he has in his criticism of de Gaulle. The magnitude of the General's legend, his intellectuality, and his integrity seem precisely what has broken many editorial hopes and hearts.

The great disappointment to the man in the street and the woman in the kitchen was that de Gaulle did not say a word in his speech about butter. The last time the French around here had butter was Christmas Eve. The food situation is frightful. Parisians are already talking about "going down into the streets," which is their phrase for rioting, and they might really do it if they had enough physical and mental energy. The four years of Nazi occupation produced a general torpor. This torpor has plainly affected the Food Ministry, which has just promised more meat next July. Parisians can now recite by heart the number of bombed bridges still unrepaired and the number of trains and trucks being used to feed the armies at the battlefronts. The statistics they would like some news on are the twenty-five thousand tons of black-market butter and the ten million francs of black-market profit that changed hands in the past six months; they would like to hear why the bombed bridges hold up legal comestibles while the black-market trucks cross the rivers as if by miracles. The French realize that the black-market butter, even if diverted to legal channels, would not spread one slice of bread per citizen per day, but what they want, and quickly, is some symbolic act of justice, invention, efficacy, and administrative gumption.

Butter has taken on political qualities. By New Year's Day certain butterless French were already muttering that they had had more food when the country was occupied by the Germans than they had had under the Allies. This slur was attributed to the ten thousand Fifth Columnists that were rumored to have been filtered back into Paris by the Germans. It is now rather generally admitted that the French have less of everything under the Allies, that the little France has she manages badly, and that there are forty thousand Fifth Columnists in Paris. Friction between the Americans and the French has been steadily mounting. The worst of it is that there is always a great deal of truth in every unpleasant claim made by either side. And it is no lie that the occupiers of a country are never popular, either those who want to stay forever or those who are dying to get back home. Good Americans here criticize those French who think their country can remain great no matter how picayunely they find it convenient to act. Good Frenchmen criticize those American Army officers who knowingly frequent the same fashionable salons the Germans frequented. One of the reasons for supposing that the war will soon be over is that all three of us old Allies have been getting along together as badly as we usually do in peace.

Two extremely curious plays that ran under the baffled Nazis have persisted into the liberation. The weightier of the pair is the Atelier Theatre's "Antigone," once by Sophocles, now put into modern dress and modern psychology by Jean Anouilh, a well-known dramatist. It is extraordinary that the Germans, with their power complex, should have permitted the presentation of this power classic which for two thousand years has been proving that individual rebellion is a good idea. The fillip of anachronism which modern dress and mode always give to the classics is especially thought-provoking in "Antigone." The two best scenes in the Anouilh version would astonish Sophocles; they feature fascinating conversations between Œdipus's daughter and a police detective. The second, far more original occupation piece is "Huis Clos," now about to reopen at the Vieux-Colombier. The title is a legal term meaning "behind closed doors;" the author is Jean-Paul Sartre, now in America. Sartre is certainly the most creative stylist to emerge during the occupation. His astonishing play is set in Hell, which in this case is a cheap hotel room. One of this hell's special tortures is a lack of mirrors for two women, the unhappier of whom is a throwback to Sappho. The general torment is that each damned soul depends on the others for proof of his or her eternal infernal individuality. Sartre is a member of the Café de Flore literary group, which has replaced the Deux Magots as the Left Bank intelligentsia center.

The Flore is the only café which looks like, though it does not taste like, old times: male and female intellectuals, jargon, no fresh air. And now no alcohol. In Montmartre, a place called the Chansonnier Dix Heures is nightly packed with appreciators of its political ballad singers, who spare nobody. One night when your correspondent was there, a song entitled "The Stars and Stripes," which jeers at the American overfeeding of German prisoners, and an impiety called "Hallelujah, the Pope Is a Republican," referring to the Vatican's recent liberalisms, gave the greatest joy. You need a strong head and legs to stay up late in Montmartre now, for you have to walk home after the last Métro, which stops running at eleven-fifteen.

The most dramatic announcement this week in Paris was that Marshal Pétain's properties have just been seized by what used to be his state.

March 30

Hannibal's crossing of the Alps with elephants is the only military transportation miracle which, to historical-minded Parisians, seems comparable, in its exciting strangeness, to our Allied amphibicaerial feat of crossing the Rhine with glider scows in the sky and ocean boats assembled on the land beneath. *Le Canard Enchaîné,* the cafés' favorite satiric gazette, rushed into print with a cartoon showing a goggle-eyed Rhineland grasshopper—*"les sauterelles"* was occupied-Paris slang for the Nazis who laid waste the French land—peering in terror at the parachute-filled spring clouds and exclaiming, *"Tiens!* It's raining men!" In keeping with the extravagance of the good news, the paper-starved Paris newspapers began spreading themselves onto two tabloid-size sheets instead of the standard one tabloid-size sheet, or onto a single, old-style big page, or they added a picture section showing Nazi prisoners, hands held high, springing through the concrete dragon teeth of their invincible Siegfried Line. The Paris journals acted as though they were drunk on ink; no two headlines agreed on what town the Allied troops were closest to. It was as if the editors had shut their eyes and confidently pinned a tail on the donkey—Cassel, Münster, Nuremberg, Danzig, Vienna—*qu'importe?* To Paris, the prodigy of Allied speed—part victory, part gasoline—was breathtaking.

The German radio, once an occupied necessity, is now a joy to Parisians. Until recently, the Cologne, Stuttgart, and Vienna stations were the three biggest the Nazis operated. After the Americans had taken Cologne and the Russians had closed in on Vienna, those two stations were combined, and if you dialled one you got either—meaning neither, geographically, since doubtless the new transmitter was in heaven knows what sheltered area. Thus, on the Tuesday Cologne fell, the Cologne station was still emitting soothing German waltzes. Last week, during which the Allies advertised that they had dropped eighty thousand tons of bombs on dilapidated Germany, the *Köln-Wien* station intermittently informed whatever German listeners were still able to hear it over the deathly uproar that "there are only weak enemy fighter groups over the Reich." However, after the Allies crossed the Rhine in their Noah's Ark flotilla of so-called buffalo, alligator, and duck boats, the German radio moved rapidly from alarm to self-pity to treachery against the *Partei.* One weekend commentator moaned, "Somebody has failed at his duty. The fact that so many depots of oil, food supplies, and munitions fall intact into the enemy's hands proves it." The most treacherous statement, for which a month ago the Germans not only would have cut the commentator off the air but would have cut his voice from his shoulders, was what purported to be a quotation from the *Völkischer Beobachter,* Hitler's and the Nazi Party's own newspaper. It went as follows: "It has become very difficult to feel like a National Socialist and to profess that one is one. We have no illusions about this." From disillusion to democracy will doubtless take no more time than will be required for the *Völkischer Beobachter* editor to turn his coat.

April 19

The death of President Roosevelt caused a more personal grief among the French than the deaths of their own recent great men. On the demise of both old Clemenceau and Marshal Foch, their grief was a nationalistic, patriotic emotion, since these men, the one with his sabre-sharp tongue, the other with his sword, had saved France. The sorrow the French felt at losing Roosevelt seemed like someone's private unhappiness multiplied by millions. Friday morning, when the news was first known here, French men and women approached the groups of Americans in uniform standing on street corners and in public places and, with a mixture of formality and obvious emotion, expressed their sorrow, sometimes in French, sometimes in broken English. On the Rue Scribe, a sergeant in a jeep held up traffic while he received the condolences of two elderly French spinsters. In the Jardin des Tuileries, an American woman was stopped beneath the white-flowering chestnut trees by a French schoolboy who, with trembling voice, spoke for his father, a dead Army officer, to express his father's love for the dead President. At the outdoor flower stalls of the Place de la Madeleine, a patriarchal flower vendor gave a passing and startled paratrooper a free pink tulip, with the statement "Today they will be sending beautiful flowers for your great man. How sad." A café waitress naïvely touched the sublime when she said of his death, *"C'est ennuyeux pour toute l'humanité."*

Since the American system for filling the Presidential chair when it is left vacant by death was unknown to most French citizens, the journals here carried an official explanatory paragraph headed *"Monsieur Truman Sera Président Jusqu'en 1948"* and quoted our Constitution.

The Paris press wrote of F.D.R. with sober magnificence and sincere superlatives. Under the spirited Gallic headline *"Vive Roosevelt!,"* the *Libération-Soir* spoke of "the unjust destiny and yet the ancient grandeur of the event." *Le Monde,* in an editorial entitled *"Après Roosevelt,"* began by saying, "The great voice which directed American political destinies has been silenced, but its echo continues in French souls." In conclusion, it praised "his charm, his beautiful and great words," and said, "Let us weep for this man and hope that his wise and generous conception of the human communities remains like a light to brighten the path for all men of good will." De Gaulle's Minister of Foreign Affairs said, "It is not only appropriate but necessary to express the depth of the sadness of the government and of the French people. Roosevelt was one of the most loved and venerated men in France. He takes with him the tenderness of the French nation."

The increasing malaise, now that Roosevelt must be absent from the peace, and the unexpected return last Saturday of thousands of

French prisoners liberated from Germany, juxtaposed fear and happiness in a mélange that Parisians will probably always remember in recalling that historic weekend. On Saturday, eight thousand French male prisoners were flown back from Germany in American transport planes, which afterward tumultuously circled the city while the men were being unpacked from trucks outside the newly decorated reception center in the Gare d'Orsay. On its walls these weary men saw an astonishing series of modernistic bas-reliefs depicting their welcome return to the freedom of what explanatory signs called *"la liberté d'aimer"* and the liberty to play, to sleep, to work, to eat, to drink, and to breathe freely. Few of the prisoners, in their hasty flight from the German Army and their later flight with the Americans in the skies, had heard our sad news. When they did hear it, one thin, bitter blond Frenchman said, *"Voyez-vous.* We've come home too late."

The next day, the first contingent of women prisoners arrived by train, bringing with them as very nearly their only baggage the proofs, on their faces and their bodies and in their weakly spoken reports, of the atrocities that had been their lot and that of hundreds of thousands of others in the numerous concentration camps our armies are liberating, almost too late. These three hundred women, who came in exchange for German women held in France, were from the prison camp of Ravensbrück, in the marshes midway between Berlin and Stettin. They arrived at the Gare de Lyon at eleven in the morning and were met by a nearly speechless crowd ready with welcoming bouquets of lilacs and other spring flowers, and by General de Gaulle, who wept. As he shook hands with some wretched woman leaning from a window of the train, she suddenly screamed, *"C'est lui!,"* and pointed to her husband, standing nearby, who had not recognized her. There was a general, anguished babble of search, of finding or not finding. There was almost no joy; the emotion penetrated beyond that, to something nearer pain. Too much suffering lay behind this homecoming, and it was the suffering that showed in the women's faces and bodies.

Of the three hundred women whom the Ravensbrück *Kommandant* had selected as being able to put up the best appearance, eleven had died en route. One woman, taken from the train unconscious and placed on a litter, by chance opened her eyes just as de Gaulle's color guard marched past her with the French tricolor. She lifted an emaciated arm, pointed to the flag, and swooned again. Another woman, who still had a strong voice and an air of authority, said she had been a camp nurse. Unable to find her daughter and son-in-law in the crowd, she began shouting "Monique! Pierre!" and crying out that her son and husband had been killed fighting in the resistance and now where were those two who were all she had left? Then she sobbed weakly. One matron, six years ago renowned in Paris for her elegance, had become a bent, dazed, shabby old woman. When her smartly attired brother, who met her, said, like an automaton, "Where is your luggage?," she silently handed him what looked like a dirty black sweater fastened with safety pins around whatever small belongings were rolled inside. In a way, all the women looked alike: their faces were gray-green, with reddish-brown circles around their eyes, which seemed to see but not to take in. They were dressed like scarecrows, in what had been given them at camp, clothes taken from the dead of all nationalities. As the lilacs fell from inert hands, the flowers made a purple carpet on the platform and the perfume of the trampled flowers mixed with the stench of illness and dirt.

May 11

In Paris the war ended the way it began—with marching. It began with the French soldiers marching off to the war and it ended with the French civilians marching around into the peace. A frenzied spirit of enlistment seemed to take hold of the people of Paris the moment peace ceased being a rumor and became a fact, and by the hundreds of thousands they settled into restless ranks and began their parade. They started marching on Tuesday afternoon at three o'clock, just after the voice of General de Gaulle announced the great news over the government's street loudspeaker system, and thousands were still marching at dawn the next morning. The crowds marched in the sunshine and on into the night with the collective, wandering rhythm of masses who are not going anywhere but are feeling something which their marching together expresses. V-E Day here was like an occupation of Paris by Parisians. They streamed out onto their city's avenues and boulevards and took possession of them, filling them from curb to curb. They paved the Champs-Elysées with their moving, serried bodies. Around the Arc de Triomphe the marchers, pouring in from the spokes of the Etoile, solidified into a dangerous, living, sculptural mass which was swayed and pushed by its own weight until the marchers, limping and dishevelled, disengaged themselves to march back down the avenues and boulevards, in the dusty, beautiful spring heat.

The babble and the shuffle of feet drowned out the sound of the stentorian church bells that clanged for peace, and even the cannon firing from the Invalides and from the precincts of the Louvre were audible on the nearby Place de la Concorde only as rumbling explosions, muffled by the closer noise of feet and tongues that were never still. The people in the crowds seemed to draw nourishment from each other, and the strength to go on down one more street, up one more avenue. Except for whatever food they may have brought from home and carried in their pockets, the marching masses lived on air and emotion. Restaurants were closed, aperitifs were scarce, beer was feeble. Peace and spring found the Parisians as badly victualled as they had been during the war and the winter, but now no one thought or cared about being hungry. All anyone cared about was to keep moving, to keep shouting, to keep singing snatches of the "Marseillaise"—"*Le jour de gloire est arrivé . . . marchons, marchons.*"

By midnight, on side streets near the Rond Point, the crowds had thinned enough for long lines of boys and girls, arms high and holding hands, to run, to crack the whip, to curvet and stretch out in antics, like long lines of noctambulistic paper dolls. Youth dominated the day and the night. It was youth that shouted with delight at the occasional skyrockets which cometed through the night sky, and it was youth, earlier in the day, that had screamed with joy at the giant American planes which roared above the tops of the chestnut trees on the Champs-Elysées and flirted deafeningly around the Obelisk. It was young Paris that was on foot and underfoot. It was the new postwar generation, running free and mixed on the streets, celebrating peace with a fine freedom which their parents, young in 1918, had certainly not known. And those Frenchmen, with their old ribbons and medals, who had fought through the last war were the quietest, the least free-moving of the Parisians among the street crowds during this week's peace. As one of these tired, middle-aged French veterans said, "That great world insomnia which is war has come to an end once again."

France and the rest of Europe are tired to death of death, and of destruction. Much of the comfort which should have arrived automatically with the peace has been lost in the news of the German concentration camps, which, arriving near the end of the war, suddenly became the most important news of all its nearly six years of conquests, defeats, campaigns, and final victories. The stench of human wreckage in which the Nazi regime finally sank down to defeat has been the most shocking fact of modern times. Only dreadful natural phenomena like the earthquake in Lisbon in the time of Voltaire or the gases and ashes of Pompeii in the time of Pliny the Younger seem to have so horrified the world. The shooting is over, but the concentration-camp prisoners are now returning, or are being advertised for by families still waiting in vain at railway stations with diminishing hope. In *Le Monde,* under the brief heading "RECHERCHE," an advertisement has just been run that said, "Any persons having known Mmes. Marie-Louise Roure and Adrienne Baumer, deportees at the camp of Ravensbrück, and Louis Baumer, deported from the camp at Compiègne in June, 1944, are begged to furnish their information to Monsieur Remy Roure." The Nazi program of non-military destruction has not been affected by the peace; it is still continuing in the bodies of the camp prisoners now home but still sick or still crazy or still mutilated or still stone-deaf from blows on the head or still malodorous with running sores.

The vigilant resistance newspaper *Le Franc-Tireur* has demanded that returned prisoners who are strong enough be selected to serve as jurymen in the still to come trials of collaborationists, of which the trial of Marshal Pétain, in June, will be the climax. The same paper has also had the disagreeable courage to say what thousands of prisoners' families know is the truth—that the Ministry of Prisoners and Deportees functions, as it has functioned from the start, with inefficiency and confusion. Many prisoners have arrived without being met by officials, or even by their unnotified families, at the railway stations; the Ministry has not provided trucks, or even ambulances, for transportation; and the pitiful thousand francs each prisoner receives as a government gift on returning home is doled out by an office which sportily observes *La semaine anglaise* and so is closed Saturdays to penniless prisoners who for years have lost all count of money, let alone time. Furthermore, the wretched Spanish Republican prisoners who were plucked out of prewar French concentration camps by the Nazis and sent off to Germany are, on their return to La Belle France, being popped back into the same old French camps again.

Some of the best and most highly educated brains of France, after imprisonment in Germany, are now being returned, ripened by isolation, suffering, hope, and misinformation. When a questionnaire was submitted to a select group of such men a few days ago, three trenchant reactions turned up repeatedly: No. 1, since they, the prisoners, had been informed in Germany that France was overrun by famine, anarchy, and civil war, they were relieved to find France merely suffering as usual from political party wrangles at the expense of national unity; No. 2, it was their opinion, formed in prison camp, that fear and lack of faith are what push any nation into self-destructive attitudes, such as Munich and Collaboration; and No. 3, it was their opinion, formed since leaving prison, that the post-liberation Frenchmen are "too inclined to easygoingness, immorality, and selfishness."

None of the Paris newspapers thought of the great and simple headline "VICTORY," which both the Paris *Herald Tribune* and the *Stars & Stripes* used on their front pages on the historic Tuesday, May 8th. Most of the French dailies described Germany rather than the Allies in the favorite headline, which proclaimed, "L'ALLE-MAGNE A CAPITULE." General de Gaulle's paper, *Les Nouvelles du Matin,* gave over its front page to a fancy drawing of a female figure symbolic of France, with wings, laurels, Allied flags, and the headline "LA GUERRE EST FINIE." *France-Soir's* headline was a great white "JOUR-V" on a deep black band. The Socialist *Populaire* used as its headline "LE REICH NAZI ABBATU." The Communist *Humanité* had no Tuesday headline at all, since it ignored the victory that day but celebrated it Wednesday, Moscow's V Day, with a seven-tier headline—or more headlines than anyone in Paris had ever seen before —which declared (we translate), "VICTORY! ACCLAIMED ALL DAY YES-TERDAY BY THE PEOPLE OF PARIS IN CELEBRATING THE MILITARY TRIUMPH OF THE UNITED NATIONS OVER HITLERISM. IN INNUMERABLE PARADES, IN IMPROVISED MASS MEETINGS, EVERYWHERE WAS DECLARED THE WISH TO CRUSH THE RESIDUE OF HITLERISM. FROM ONE END OF THE LAND TO THE OTHER, BUT ONE UNANIMOUS CRY: PETAIN TO THE GALLOWS! VIVE LA FRANCE! VIVE LA REPUBLIQUE!" The incorrigible *Canard Enchaîné,* Paris's favorite satiric gazette, celebrated the victory by publishing a cartoon of Hitler, dead and gone to heaven, where he starts the New Order by pinning a Jewish star of David on God's chest.

At half-past-ten Monday night, in the Allied Servicemen's Club in the Grand Hotel, an orchestra leader whose men had just finished playing and singing "And the tears flowed like wine" suddenly shouted to the dancers that the war was over, that tomorrow was V Day in Europe. A wild groan of joy went up from the men in uniform. Then they began capering extravagantly, their girls in their arms, to the tune of "Hallelujah." Afterward, tumbling out onto the boulevards, they started spreading the news to all Joes and to anyone French who would listen. But that night it was only the Americans, the optimists from across the Atlantic, who believed the good news about Europe.

Next morning, though, the French papers announced that at 3 p.m. the great news would be officially true. At that hour all Paris was dressed in its not very fine best and standing on its handsome boulevards waiting in the sun for the "Marseillaise" and the pronouncement of its leader de Gaulle. And then his for once impassioned, excited voice proclaimed victory, which had come at last.

Europe's Five and a Half Years' War is all over but the peace.

May 24

Now that not only Paris but all Europe is liberated, it seems reasonable to view Paris as what is left, after another plague of German war, of the city that was once the civilized, intellectual capital of so-called French Europe. Paris is not gay; it is restless, anxious, cantankerous, and probably convalescent. The theatres are packed; the shops and groceries are still empty. Parisians either have money to throw away and little to throw it away on, or they have next to nothing and manage to live on it. There are three classes: first, the formerly comfortable *rentier* class, with incomes that now buy a tenth of what they did; second, the working class, which, with the highest wages in French history, is now striking all over France for increased wages to meet increasing prices (in the northern coal mines, however, where production had fallen off anyway because the miners were on a vegetable diet, the strikes are really to get meat); and third, the inflation-money class, both new rich and old rich, who are part of the financial phenomenon of accruing cheap millions and scarce goods the way floating leaves and logs are part of the phenomenon of a river in flood. Class No. 3 looks, so far, like the class of survivors.

The first new American movie to reach film-famished Paris was "Mrs. Miniver," whose sweetness and light were given a peculiar twist by the fact that the opening night was a gala charity affair to raise money for the sick, sad women returning from the concentration camp of Ravensbrück. The only other big American movie that has been seen since the liberation is Charlie Chaplin's "The Dictator," which Paris, after four years of one, did not find funny. Maurice Chevalier has been packing the midtown ABC, which is again offering, as before the war, picked night club talent on a music-hall basis. Ciro's has been taken over as a private night club by French officers, who, since the Americans and British have taken over everything else, had to have some place to go. Ciro's has the best entertainment in town, including old, familiar faces like the fabulous Charles Trénet and the fascinating Edith Piaf and her tough songs.

July 26

Certainly it is incorrect to call Marshal Pétain's trial for treason, in the Palais de Justice, one of the greatest trials in history. The trial is turning out to be merely one of the greatest lessons in history, pedagogically speaking, the French people have had since June, 1940. It was then that there began the particular chapter of French history whose details most French citizens are only now learning about, five years later. The lesson is being recited ad lib by expert witnesses like ex-Premiers Paul Reynaud and Edouard Daladier and ex-President Albert Lebrun in the court sessions, and next morning the French newspapers are giving the text to the ignorant citizens. Each sad passage is headlined in a manner that history books perhaps should copy. "Audience Instructed" was one of *Paris-Presse's* opening headlines, followed by a subhead explaining that "the disastrous events of the year '40 are still not fully known except by a small number of initiates." In the same didactic spirit, *La Voix de Paris* ran the headline *"Révélation Sensationelle de Daladier,"* plus a subhead that gave Daladier's government's figures on

France's 1939 air force, and then, in the article that followed, went into the Marshal's previous request for a reduction of armament appropriations, his lack of interest in those military modernities like tanks and anti-tank guns, his peacetime domination of the French Army chiefs, and so on. *Le Monde* featured Reynaud's revelations about those last, complicated, disorderly days of the Republic in Bordeaux, under a subhead saying "Prisoner Pétain's Plot to Seize Power Via an Armistice." Other journals have featured the fact that, in Bordeaux, Pétain had ready in his pocket, like an extra handkerchief, a tentative list of names for his prospective autocratic cabinet. The collaborating policy arranged at the Montoire meeting between Hitler and Pétain also got a big front-page play. Even the Battle of Verdun was slipped into the court lesson and into the reporting of it, together with the little-known fact that, according to General Joffre, its hero was not the pallid, handsome, prisoner Marshal, alive and in his armchair before the jurors, but a minor, now-dead general named Nivelle. And, finally, the French armistice of 1940 hit the front page, after the Marshal's unattractive, sour, stammering, black-robed chief defense advocate, the *bâtonnier* Payen, angrily inquired if anyone present in court so much as knew the armistice terms, and after the charming, smiling Presiding Judge Mongibeaux leaned down from his bench, rather like an elegant host catering to his guest's whims, to ask if perchance there was a copy of the armistice in the house. At that point, the terrifying old Prosecutor of the Republic Mornet, who wants Pétain's head and has said so, leaped up from his box to recite by heart those armistice clauses that gave to the Nazis the right to occupy two-thirds of France. "That's unfair!" the defense screamed. "You've no right to begin with the worst part." "I have a right to begin where I choose and my good memory dictates," the prosecution snarled back. The red and black robes recalled the face cards in the trial scene of "Alice in Wonderland," but in the Paris trial scene there was no gentle humor, no awakening from a dream.

The general outline of French history from 1940 to today, most of the details and anecdotes, and certainly more documentation than has yet been cited in court have long been familiar to all the American journalists who have been sitting on their press bench with pencils idle in the heat and, listening, have refreshed their memories. There are four reasons why the French did not until now know their own facts. In June, 1940, when France was falling, people and newspapers were fleeing Paris, as if to avoid the country's collapsing on them physically. In the confusion, the June events were only fragmentarily or inaccurately known. After Pétain settled firmly into power in Vichy, he was at first too beloved for anyone to tell the truth and later it became too dangerous to tell. At the time of the Riom trial, at which some of the witnesses in today's court were then prisoners, audaciously giving the same news of France's fall (as a means of expressing faith in America), the excellent underground Riom reports were sent out to the United States but had little circulation in France. And after the liberation last summer, when the truth about France's past could have been told, the exalted French were interested only in her future. Anyhow, there was a paper shortage, which would have prevented printing this vast stale news even if anyone had been interested in doing it. Now, during these long, hot hours of Pétain's trial, the old news is datelined afresh; it is as good as, it is worse than, new. The French absorb it hungrily or cynically, as proof of this or as negation of that. It embodies everything that French democracy is now facing.

There are already three important rumors about the way the trial will end, and all are untrue. The first two are that the French penal code forbids the execution of a man over seventy and that the special legal weight of a high court such as the one Pétain is being tried in makes impossible a reprieve of the death sentence. The third false rumor is that General de Gaulle is against the death sentence. Many Parisians say that they would be satisfied if the old man's buttons were cut away—that is, if he suffered military degradation, a punishment perhaps worse than death to him—since they are sick to the depth of their souls of eternally hearing about Le Maréchal.

Under the complicated but fair system by which the French Ministry of Information parcelled out the tickets to the courtroom, which seats only about six hundred, your correspondent's first ticket was for Tuesday, the second day of the trial. The initial visual impression was not of the white-faced Marshal but of the picturesque, scarlet-colored opening of the court that had assembled to try him. This began with the filing in of the five red-robed judges, black velvet hats in hand, and ended with the entrance, loudly announced, of *Monsieur le Procureur Général*. Preceded by a major domo (a sort of kindly, bald court butler in a worn dress suit), Prosecutor Mornet rushed into the room, past the red-robed Bench, and toward his private little pulpit—a torso bent almost horizontal with age, eagerness, and speed, his pointed gray beard and peaked nose leading, his blood-colored robe trailing behind like a sanguine silhouette from a sketchbook of Goya. One's mental impressions of the court accumulated more slowly. To an American, whose judges are supposed to be impartial arbiters, the active, witty, worldly, and openly anti-Pétain comments of handsome Presiding Judge Mongibeaux seemed a strange foreign novelty. It seemed equally odd that once a deputy juror interrupted witness Reynaud to back him up in some statement; that two resistance jurors jumped to their feet to protest angrily against what they called Pétain's *bâtonnier's* insults to their cause; that the same *bâtonnier* accused Reynaud on the stand of calumny, then of "remaking his political virginity"; that both Prime Ministers were allowed to quote what somebody said somebody else had said; that both were allowed to talk by the hour, often of events long before their political time; and, above all, that they were allowed to interpret their own part, their nation's part, the Marshal's part in any event whatever—that they were, indeed, encouraged to stand, with the eyes and ears of the world watching and listening, while they mixed memoirs and politics into their lecture on, and version of, recent French history. It would appear that this peculiar academic treatment, in this particular place and court, was clearly the only procedure by which the final verdict could be reached. As Daladier declared on the stand, with the brutal candor of an *Homme d'Etat* newly sworn for that afternoon to the whole truth, "There is no connection between justice and politics."

During Reynaud's endless testimony that day, he spoke, among a thousand other things, of the fact that the poet Paul Valéry, once Pétain's friend, had delivered the address of welcome when the Marshal was finally received into the French Academy. Tuesday, just before midnight, the body of Valéry, who had died bravely of cancer, was escorted, to the beat of muffled military drums, through dark streets to a flambeaux-lighted outdoor catafalque, where it lay, keeping its own last nocturnal vigil over that great view of Paris between the Trocadéro hill and the Tour Eiffel. Valéry was the greatest living French poet, as Pétain had been the greatest living French Marshal; they were of the same vintage and at one time they had been equally appreciated, each for his special flavor. While in Vichy, Pétain dismissed Valéry from his post in that curious institution of scholarly Paris exiles, the Centre Universitaire Méditerranéen of Nice. De Gaulle later reinstated him. In 1943, the elderly poet joined the clandestine littérateurs of the resistant Comité National des Ecrivains. On that Tuesday night, there were only a handful of us waiting on the curb as, in a sudden red flare of Bengal lights, like an apotheosis, the dead poet was carried from the Church of St.-Honoré-d'Eylau. Two once-great old men of France had just been lost forever, and a period of French history had indeed come to its end.

August 3

It is extraordinary that the treason trial of Marshal Philippe Pétain reached its sixth day in the Palais de Justice before it bumped into a few completely objective facts. Up to then the trial had been coasting along on the airy, wary personal opinions of some of the most famous and least disinterested politicians in France—those who had been running the country during the already culpable nineteen-thirties, before the Marshal took over at Vichy. The little-known, fact-giving witness was General Paul Doyen, who had been the French president of the Wiesbaden Armistice Commission until he was sacked, early in the occupation, by the Vichy crowd for telling them that they not only should but could resist German post-armistice pressure. The important facts, which, the General declared, made "a recitation painful to French ears," featured Germany's unprotested illegal annexation of Alsace-Lorraine, which, though it happened in August, 1940, was still news to Paris in July, 1945. They also included the ensuing Nazi demand for all French war material in North Africa, as feed for Desert Rat Rommel's campaign against Wavell in Tripoli, and the Vichy Government's promise to comply, which General Weygand was able to sabotage by slowdowns. And they detailed the subsequent German demand for the Levant airfields, plus the port of Bizerte, and Vichy's compliance, which by this time no one, either military or civil, could slow down. The climax of General Doyen's declarations was that the Marshal, after kicking out Dauphin Pierre Laval in early December of 1940, "thereafter had at Vichy a free road to defend what could be defended of France" against the Germans, but that "of his own choice he had elected otherwise."

The importance of Doyen's testimony (aside from his revelation that the elegant Marshal, during the first Christmas at Vichy, which was to have been so amical, referred to Laval as "a dung heap") lay in the fact that it was indeed fact, based on the Wiesbaden documents, and that it dealt with the Pétain of the occupation, surely the period in which the charge of treason will have to be localized if it is to be made truly legal. One of the weaknesses of the prosecution up to then had been that its political witnesses had droned on about the prewar anti-republicanism of the Marshal, as if that in itself was a crime during the Third Republic, which unfortunately it was not. The prosecution's other great weakness had been that it had achieved only a single, oblique reference to that historic telegram from President Roosevelt which requested Pétain's aid for our invading Allied troops in North Africa, and to the Marshal's reply that he would resist any invasion, and to his not resisting the Germans when they invaded his unoccupied France a few days later, on November 11, 1942. To the French populace, confused by its old mixed love and hate for the Marshal and addled by the current tarradiddles in court, that November 11th has become Treachery Day just as significantly as July 14th became Independence Day.

One thing the populace has certainly disagreed with the prosecution about is the delay in calling to the stand some of the resistance families that lost men and women in the struggle, tortured or shot by Pétain's militia or sent by them to die more slowly, scarred and starved, in Nazi concentration camps. One such resistance relict presented herself early this week on the sidewalk before the Palais de Justice gate. She was a middle-aged country mother, sweating in her heavy mourning in the afternoon heat. She told the Palais guard that one of her sons had been beaten to death by Vichy legionnaires, that the other had perished in the concentration camp of Nordhausen, and that she wanted to speak to the Marshal. When an American reporter urged the guard to let her enter (so-called "spontaneous witnesses" are admitted to this strange high court), the guard refused, saying, "She might assassinate him. Then his death would be my fault." Later that same afternoon, Colonel Charles Chaudrey, the resistance hero celebrated under the *nom de guerre* of Grégoire, did indeed speak to Pétain in court. Chaudrey pointed an accusing finger at Pétain and cried, "If there remains in him even a little love for France, let him, before he pays with his life, fall to his knees upon her land and ask its pardon!"

* * *

For the American press, the most ticklish moment in the trial came during the reading of the pro-Pétain letter from Admiral Leahy, our former ambassador to Vichy. To this communication, which interrupted one of the many installments of Reynaud's uncomplimentary evidence against the Marshal (the defense had shrewdly cut in with Leahy's praise, to prove Reynaud wrong), the sly little mouse-faced politician came back with a gallant pun. "Ambassadors are polite enough to forget the truth. *C'est la courte mémoire qui fait la courtoisie diplomatique.*" The Parisian press seemed almost bashful about acknowledging the existence of the Leahy note. Both *France-Soir* and *Ce Soir* lied about it politely. The first briefly described it as "a rather reticent letter, which could cut both ways." The second was more thorough, explaining, truthfully enough, that the defense considered that it had had a big stroke of luck in the Leahy document but that all Leahy really had said was that Pétain should have resisted the Germans more, even if France suffered more. Only *Le Monde* actually quoted the Leahy letter intact, but it did so without comment, which its conservative readers could easily make for themselves.

What with the touchiness of Franco-American relations and the French people's angry memory of what they called our State Department's misleading backing of Vichy in the early days, it is just as well that the populace here never reads *Le Monde* and so missed Leahy's choicer paragraphs, translated into lovely French. Example: "*J'avais alors et j'ai maintenant la conviction que votre but principal était le bien et la protection du peuple abandonné de France . . . Avec l'expression de mes sentiments personnels et avec les vœux que votre activité dans la période d'occupation par l'ennemi puisse être évaluée à sa juste valeur par le peuple français, je demeure très sincèrement à vous,* Leahy."

Pétain was in no position to have the best lawyers in the land come running to his help, but his trio of defenders still seems peculiarly, unnecessarily poor, lacking tact, teamwork, grace, and, with one exception, brains. His chief, *bâtonnier* Payen, has a scattered, jerky mind, and an unfortunate tic which makes his mouth pop open and shut like a bullfrog's. Pétain's second man, *maître* Lemaire, raises unimportant legal points in the loud, declamatory voice of the Comédie-Française. The youngest Pétain defender, *maître* Isorni, has brains, but they don't seem at their best at the bar, where his bad habit of continually interrupting everybody seems to distress even his client, as Judge Mongibeaux classically and courteously calls old Pétain. All that can be said for the defense is that no one, at any rate, has yet accused the pro-Pétain trio of having had anything to do with the resistance. Yet all the court's major anti-Pétainist witnesses have had to admit, under cross-examination, that once they were pro-Pétain, in that they trusted what is now constantly called his prestige and authority. As Léon Blum thoughtfully told the court, "There is a Pétain mystery." The judges now trying Pétain for his life once perforce swore allegiance to him as their Unoccupied France chief of state. Even old Prosecutor Mornet had to admit, after first denying the fact with an indignant screech of "Infamy!," that he had accepted an invitation to sit on Pétain's scandalous Riom trial bench, though he later refused to take his seat. And even the two juries, one chosen from the resistance, the other from those rare parliamentarians who in Bordeaux had voted against giving the Republic to the old, autocratic Marshal, even the juries—like those of the French Revolution, out to get heads—are packed and proud of it. So far, the trial has tarnished everyone it has touched. An atmosphere in which all the men seem faintly fallible and all the methods slightly illegal has established itself as the natural air to breathe in the small, stuffed, shoddy courtroom.

Of all the testimony of the big political figures, what Léon Blum had to say against Pétain was the most intellectual, clear (if complex), and unequivocal. All the politicians were asked by the Presiding Judge if they thought Pétain had committed treason and what they thought treason consisted of. Blum's definition was best: "An absence of moral confidence was the base of the Vichy Government, and that is treason. Treason is the act of selling out."

Taking advantage of one of the admirable, odd privileges given to them in this high court, entitling them to insist upon hearing from anyone they chose, the resistance jurors insisted on summoning the newly returned Pierre Laval. Laval's name is a well-known palindrome—as the French say, it reads the same forward and backward, so how can you tell if he's coming or going—and Parisians wagered that Laval, being too dangerous to both the defense and the prosecution, would not be called to talk. He was called, and he talked. He entered the courtroom as if its complete silence were dangerous to him. He carried his gray hat and his worn brown briefcase in his hand, in confusion and in a low voice asked where the witness stand was, and, when he saw that it was only a carved, cane-seated parlor chair, deposited his hat and briefcase on the seat and stood erect. He was at first unrecognizable. The fat of his face is now gone. His oily, Moorish hair is now dry and gray and his mustache is the color of tobacco juice. His crooked, stained teeth make a dark, cavernous background for his large lips. Until he spoke, he was some other man who had been announced as the infamous Pierre Laval. The buxom dandy of other days had turned into an unprosperous stranger, a worn, brown-skinned, worried-looking country fellow, with watchful, inelegant, careful gestures. Above his habitual white string tie, his large, low white collar ruffled around his thin neck, which was hung with loose wattles like those of a brunette turkey cock. His rumpled, gray-and-white-striped suit was so large for his frame that it looked borrowed. The moment he started speaking, the interesting phenomenon ended. His voice, his vocabulary, his alert and vulgar mentation were the same as ever.

For thirty-nine minutes by the clock, Laval talked without a stop and without ever answering Judge Mongibeaux's opening inquiry, "When did your political relations with Marshal Pétain begin?" The only insulting thing Laval said about the Marshal was that Pétain had never known anything about politics. Then, like any successful black-market businessman telling another how he got his start, Laval confided, "Politics has to be learned. It's not something you just catch on to. It has to be studied to be good. You've got to learn it all the way up. Now, me, I began at the bottom and took on every job on the way to the top." He also recited a confidence about the Duke of Windsor, then the Prince of Wales, with whom he must have had, years ago, a curious conversation about the problem of Mussolini and the Abyssinian War. "'You ask your father about it,' I kept telling the Prince," said Laval. "'You go ask your father the King and he will tell you.' But the Prince, he kept saying that his father was a king and so wasn't a politician. Pétain was like that. War and Les Invalides were his element." Inflexible, wax-faced, motionless as usual, Marshal Pétain, from his worn black-leather easy chair, watched his former Dauphin's back. The two men did not speak, upon finding themselves together in the same high court. But once, when Laval turned, with the Vichy Dauphin smile, to say something nice about the old man, the Vichy chief of state lifted his white hand in an almost imperceptible contemptuous, familiar gesture of greeting. It was their unofficial *rencontre* at their treason trial.

August 17

The Pétain trial ended the day after the end of that World War in which Vichy had played its paltry part. The B.B.C., in London, erroneously reported riots before the Palais de Justice. Actually, Paris was calm while waiting for the verdict and deeply satisfied when it came. The French people felt that out of the lengthy legal complexities which for twenty-one days had stretched justice almost to the breaking point, out of the parade of overimportant political witnesses and too obscure generals, out of all the documents read aloud and the phrases of praise and vilification that were cited, out of all that issued from the mouths of the old Marshal's defenders and his prosecutor, it was what Marshal Pétain himself had said or written during his period of power at Vichy, and the kind of justice he had established there, that in court condemned him. It is certain that never before has an illustrious and complex old figure on whom so many millions of words have been spent been condemned on so few, and all those his own. In Prosecutor Mornet's *réquisitoire,* on the eighteenth day of the trial, there were a half dozen outstanding determinants against Pétain, among them excerpts from his radio pronunciamentos in the early Vichy days: "All our miseries have come from the Republic. . . . The responsibility for our defeat lies with the democratic political regime of France. . . . A strong state is what we wish to erect upon the ruins of the old, which fell more under the weight of its own errors than under the blows of the enemy. . . . You have only one France, which I incarnate." His letters to Hitler, only recently made known and therefore especially shocking to the public, told even more terribly against him. On the anniversary of his only meeting with the Führer, at Montoire, where Pétain offered the policy of collaboration, he solemnly wrote Hitler, "France preserves the memory of your noble gesture." And after the tragic British failure in the Dieppe Commando raid, he wrote Hitler, "I thank you for having cleansed French soil." Later, under Nazi pressure for greater conformance to the anti-Jewish laws, the Marshal fell so low as to write von Ribbentrop, "In future, all modifications of our laws will be submitted to your approbation." As the prosecution brought out—and this was the only moment during the trial when the Marshal's face looked like a marble mask of shame—a hundred and twenty thousand Jews were thereafter deported from France and only fifteen hundred lived to return.

On the nineteenth day, the defense's supposed trump card was played and came to nothing. The trump card was the theory of the Marshal's *double jeu*—though he had openly cabled President Roosevelt, "I have learned with stupefaction of your troops' aggression in North Africa," he had also telegraphed in code to Admiral Darlan, ordering him to aid us. A letter from General Juin, read in court, proved that the Marshal had indeed belatedly ordered the go ahead in code but had also cancelled it the same way because he was "in relations with Hitler."

From the very first day of the trial, Pétain's haughty opening declaration that he would not deign to talk and that the special high court lacked legal authority to try him—he who had founded a no less special high court at Riom, where Daladier and Blum had had to talk and talk speedily to save their lives—told against him. The last straw was the Marshal's strange insistence that he, too, had been among "the first of the resistants," for, he asked, had he not held France for de Gaulle to recover, and what recovery could the General have effected for a France in ruins wrought by lack of German collaboration? Among the many ruins of France today are the graves of the Gaullistes whom the old man of Vichy ordered hunted down as rebels. As Blum said on the witness stand at the trial, "I never knew Pétain. *Il y a en lui un mystère*"—a puzzle. The sole great importance of the *procès Pétain,* supposedly one of the greatest trials in history, is that it tore into a half-dozen clear pieces, for those who were willing to look and learn, the legend of the *mystique autour du Maréchal,* that unrepentant old man of unrepublican France.

October 11

There was also something mysterious about the Laval trial held in the same august court: why was it the scene of such a loosing of passions? Judges, juries, lawyers, deputies, prosecution, a youthful visitor or two in the little gallery, and, above all, the accused himself, hemmed in in his narrow aisle on the crowded floor, all shouted as they undoubtedly never had before in public. It was probably sheer hatred of Laval, the physical as well as the symbolic *summum malum* of France's occupation, collaboration, and shame, that loosed the outcries against him. And there was something left in Laval that made him scream back. Physically and mentally, he completely dominated the court both when he was in it—sweating and thinking and swearing—and when he was out of it, when he took his victory with him to his cell and the court dragged on, dull and defeated. He had his judges, prosecutor, and attackers beaten three different ways; he, the single traitor, was more intelligent, wise, and intuitive than all of them rolled together. His superior brain guided even his cheapest gibes. As a former lawyer and Minister of Justice, he knew the law better than the Bench seemed to. As a former deputy, he knew all about the mentality of deputies who made up the vituperative parliamentary jury. As a politician with an amazing memory, he knew the policies he had made and recalled them verbatim while the politically ignorant court scuffled through its reference papers for information. When he started to talk, he was like quicksilver: no one could control him; his phrases, flashy or weighty, rolled in all directions through the court, separating, then coagulating, but always slipping through the fingers.

Throughout Laval's trial, the whole Paris press demanded his death, deplored the scandalous fashion in which it was being led up to, and noted each paradoxical pickle his maneuvers put the court into. The best cartoon on the trial was run in the disrespectful weekly *Le Canard Enchaîné*. It showed two troubled lawyers talking, and the caption was "Do you think he will be able to pull himself out of all this?" "Who?" "Mongibeaux, the judge."

For five full days the French press, from Left to Right, groaned when it thought about what the foreign press bench at the trial was going to say about the affair. Yet the French press was harsher toward its own countrymen than any of the four English-language Paris papers. As angry a French sample as any was one Paris editorial which summed up, "The astuteness of Pierre Laval, that old war horse quick to exploit any situation; the feminine nervousness of a judge inclined to the vapors and attacks of nerves; the senile peevishness of a prosecuting attorney already discredited for his part in those other trials; the unimaginable lack of propriety of the jurors, who have forgotten what the most elementary decency is like—these are the elements of the scandal that is dishonoring our high court. It is enormously painful to speak in these terms of the highest magistrates in France. However, it must be said because it is true."

The pessimists' morning paper, *Combat,* which rejoices in a gloomy view of things, croaked as its contribution, "The Laval trial has the prospect of turning France's stomach. Yesterday's séance was even more odious than the two before. This time it was sordidness and mediocrity at loggerheads in the midst of incoherence. How could it be otherwise when we are watching magistrates who not only cannot control the prisoner but cannot even control themselves? For four years people have dreamed of the Laval trial. We too dreamed of it. Dreamed of settling an account. But we never imagined that Laval's trial would be as comic as it was the day before yesterday, as odious as it was yesterday, and as ridiculous as it will be tomorrow. On our manner of judging Laval, we ourselves are being judged."

There were also moments of wit and sheer fun in the trial, as when Laval coolly replied to the judge, whose question he refused to answer because the judge had already imprudently answered it himself, "As the victim of a judiciary crime, Monsieur le Premier Président, I do not aspire to be also the accomplice." And there was a roar of laughter when the nervous judge, making an elegant attempt at regaining lost authority, requested the sheriff to bring in a string of witnesses and the underling returned with nothing but the explanation "There aren't any witnesses anywhere, Monsieur le Premier Président. Perhaps we forgot to order any." There was also some fretful gaiety over the occasional tardiness of the Resistance jurors in turning up for duty; many of them were busy running for deputy in the forthcoming elections. They were and are literally running—on five new tires, three new inner tubes, and three thousand litres of gasoline allotted by the government to each of the dozen or more contending political parties, while country doctors and priests often have to walk miles to save men's lives and shrive men's souls. There were three women jurors among the Resistants, all of whom were widowed by the Laval-Pétain-Darnand-Maurras militia. The best known of the trio was the Communist Mme. Gabriel-Péri, whose dark beauty was like an allegorical portrait of Grief waiting upon Justice. And it must be noted, finally, that it was the parliamentarians, not the Resistance jurors, who conducted themselves with so experienced a lack of public dignity.

* * *

Late on the third turbulent afternoon of his trial, Laval refused to return to court after some parliamentary jurors had called him foul names and promised him "twelve balls in the skin"—or execution before the dozen rifles of the firing squad—and after the judge, as Laval put it, had submitted him to "outrageous treatment," so the prisoner was forced by legal etiquette to send a formal written declaration, announcing his refusal, by a sheriff, who carried it from the prisoner's cell in the basement of the Palais to the court, upstairs. According to the press, when the note was received by the judge, it bore, legally enough, not only Laval's signature but the sheriff's postscript, "Cost—nine francs." The next morning, Laval was reported by Palais gossip to have said to his guard, "Anyhow, I wager I slept better last night than Mongibeaux." It was this sort of realistic insouciance which drove the high court crazy.

Laval's refusal to return to the court, even to hear his sentence on the fifth day, made it necessary for his defense lawyers to go down to his cell at dusk to announce the fatal news. They arrived beside his cot during one of the shutdowns of electrical current that are a daily occurrence in Paris because of the coal shortage. His cell was lighted by a small flare. When he saw their sad and shadowed faces, he gently said, or so they later declared, "Am I the one who should try to cheer you up?" Twenty-four hours later the Paris papers, no longer querulous, grimly announced that Pierre Laval was installed in one of the death cells in the suburban prison of Fresnes. He was already dressed in a drill suit, his head had been shaved, and he was wearing leg chains.

October 20

To the sound of the groans of long-unused machinery and in a flaky cloud of rust, things are beginning their slow improvement here. But as General de Gaulle recently, and harshly, said, it will take the French people a full twenty-five years—"a whole generation of furious work"—even to resuscitate France. The general opinion is that the gigantic task of the country's complete recuperation cannot conceivably be concluded until around 1975, and that it will probably be 2000 before Europe, whatever it will be like politically then, will have recovered physically from its losses in people and cities, and from its memories of its losses in beauty.

The brightest spot in the Parliamentary pre-election hodgepodge has been an odd party called the Union Monarchiste, whose candidates have consisted of an anomaly listed as a Socialist Monarchist, one female Traditionalist, one plain Monarchist, one plain Royalist, and one Young Royalist. The candidates are not relatives of the Comte de Paris, the only royal pretender family; the candidates are merely little politicians who would like to see monarchy restored in France. The Party's boulevard posters advised French voters to keep their liberties, "like the English, Swedish, Norwegians, Danish, Belgians, and Dutch, all citizens of democratic monarchies," denounced all three French Republics, and ended with an exhortation to "rise against all tyrants!" That is exactly where, a hundred and fifty-three years ago, the First Republic came in.

October 25

Never before in living memory has a bitterly contested French election made a trio of rivals so happy as the Communists, the Socialists, and the new Popular Republicans are right now. One way or another, all of them have just won in France's first national election since before the war. The world waited over the weekend, with unusually flattering attention, to see how France would choose. France chose three things, all different. Madame La Quatrième République is starting out like a woman with three hands, two Left and one Right, the Communists and Socialists being on the side closer to her newly reawakened revolutionary heart and the Popular Republicans on her other, purse-carrying side. The Sunday elections killed the Third Republic, for the people voted enthusiastically to abolish its constitution of 1875. Probably nobody but the meticulous General de Gaulle has even reread the old constitution lately; no newspaper reprinted it for final inspection; and no candidates who were against it ever quoted from it to show what was wrong. The single sad post-election cartoon showed Marianne, symbolic figure of the now late Third Republic, gazing pitifully at the baskets of votes against her that were being brought to her deathbed, from which she wailed, "And not one love letter in the lot!"

During the campaign, tomatoes were twice thrown at ex-Prime Minister Edouard Daladier, the diehard Radical Socialist ex-chief, by country audiences on whom he was pressing his candidacy for reëlection. Léon Blum was not only still too fatigued by his Buchenwald captivity but also too wise to put himself in Daladier's position. He ran his Socialist party—into second place—but he didn't make the political error of running himself for office. The French citizen was sick of the old, bitterly disappointing, though perhaps well-meaning public faces left over from before the war. As de Gaulle had pointedly and brutally said, France wants the "new and reasonable."

If the voters acted calmly, those they were voting for did not. The harshest and most significant ill-feeling was that which developed between the Communists and Socialists, who were supposed to be blood brothers. On their ability or inability to make a majority together hangs France's immediate future, which will consist of either severe structural reforms or else a bourgeois sigh of relief. The Socialists accused the Communists of all the old types of election malpractice—stealing ballots, menacing Socialist Party members, and, above all, dropping Communist leaflets on Limoges from a couple of airplanes, though private flights over France are still forbidden—and where the gas for the planes came from the Socialists couldn't guess, unless it came from de Gaulle's Communist Minister of Air. To all this the Communists offered their usual disciplined, hermetic silence. To the bourgeoisie, however, they did make a startling reply by letting a great dead artist speak for them. In their victory edition of *L'Humanité,* among the snapshots of their winning candidates, they put Daumier's lithograph of a Gulliver-sized figure of France sweeping out with a broom the bedizened, Lilliputian bourgeois figures cluttering the French landscape like fat, fancy little insects. This is the true meaning of the election. The main plank in Blum's campaign platform was his condemnation of the French bourgeoisie as the lost, inadequate political directors of France. The Socialists want not revolution but radical change. The Communists want changes which will be revolutionary, including, to begin with, the nationalization of credit, which means "Down with the Banque de France." The rightist Popular Republicans want what they carefully call reasonable change. It is their last chance.

Whatever the new political government turns out to be, the government itself has already turned into a postwar monster. As fast as the Allies hand back the Paris hotels which they requisitioned for war, the swelling French state absorbs them as offices for peace.

Ministers mingle with chambermaids in the service of *étatisme,* and fifty thousand Parisians cannot find a place to live. Before the war, seven hundred thousand French worked for the state as functionaries; now two million are busy at it. Louis XIV said, *"L'état, c'est moi."* Now the state, unfortunately, is beginning to be everybody in all France.

December 5

Because of the worst drought Europe has known in a hundred and fifty years, Paris electricity, which has been trying to do on limited water power what it failed to do on no coal, is now rationed more severely than it was last winter. The working hours of factories not producing necessities have been cut nearly in half, and so has the pay of their workers. Two hours a day are clipped from the time department stores are allowed to burn their lights, and two hours from the salesclerks' pay. Misery is again adding itself to penury. The Métro service has been reduced one hour daily, and at night the street lights, which for a brief moment after the war comforted everyone again, are once more dimmed. Householders who had no coal last year now have a seven-day supply for each person in the family for the entire winter and are saving it for the week when they fall sick. Everybody's five-year-old chilblains have started up again. Smart, chilblained Parisiennes beg transatlantic travellers for American vaseline the way smart New Yorkers used to beg for Paris perfume. There was supposed to be a small ration of coffee beans, but what has shown up instead is a stock of old powdered coffee. However, at least French women have just been given their first cigarette ration since Marshal Pétain's virile Vichy decision that only men should smoke. Many French women do not want to smoke anyway, and their hundreds of thousands of accumulating packages of Gauloises are already weakening our soldiers' black market in American brands.

This winter's rationed food for the French costs double last winter's, but necessities are less necessary and cheaper on the black market. For the first time since the Nazis tried to call in all French firearms, game is legally on sale for the approaching holidays. A dozen dainty, dead skylarks, worth fifty francs before the war, sell at only a hundred and eighty now, and a cock pheasant fetches only around six hundred. Live game is plentiful; it is the cartridges to kill it with that are rare. One literary Parisienne, browsing in her attic among prewar books, found two boxes of prewar cartridges. For twenty shells, a neighborly crack shot brought her back in trade two pheasants, a kilo of country butter, and a roast of veal. Gas stations are open again and rejoice the motoring eye, but gas is short. Because trucks lack gas, the monthly wine ration, which had been increased to two litres a person, has been reduced to one. International trains are beginning once more to creep across Europe, whose landscape is too old-fashioned, too cruelly beset by barrier mountains, by mist-producing boundary rivers, and by windy weather to be hospitable in winter to modern man in the air. Even the wing-minded American Army is taking to slow trains. From Paris to the Nuremberg trials is now a thirty-hour, sit-up train jog. The deadly necessity of getting planes into the freezing clouds to fight is now finished. In the depressing fog of the first winter's peace in Europe, man has come down to earth again and is moving about slowly.

Two winters ago, Paris pre-Christmas shop windows were bare. Last winter, the little they showed was not for sale. This winter they are full, in special neighborhoods, for the special, full purses of the nouveaux-riches. Behind glass, in the Faubourg-St.-Honoré, luxury is purchasable, and without ration tickets—soft sweaters for both sexes, fur-lined gloves, or dreamy silk nightgowns at nightmare prices. In the windows of the Au Printemps department store, catercorner from our most popular Army PX, the traditional mechanical-doll tableaux and the live clowns, all back in action for the first time in five years, have necessitated the same old traditional line of police-men and ropes to handle this season's mobs of dazed French children, the smallest of whom stare silently at the first toy of their lives. Only those who have seen toys in their pretty past cry out with recognition. Paris is not the same and probably never will be, but under disheartening conditions it is making its first effort. After all, Christmas is a civilized fête, however little it has now to do with peace on earth.

Albert Camus's fashionable philosophic play "Caligula" opened a month ago at the quaint neighborhood Théâtre Herbertot. Camus was formerly editor of *Combat,* the most intelligent, stimulating newcomer that the resistance has given to French journalism. He has abandoned his fascinating editorializing about France and against the United States in order to write plays. "Caligula" is a drama about what one Roman did in Rome, and it illustrates the new, modish philosophy of existentialism developed by that other talented ex-*Combat* journalist, Jean-Paul Sartre. As nearly as can be made out by dullards who would have thought that an important new French philosophy must be founded on something more than a "disgust for humanity," Sartre's form of existentialism is, indeed, founded on a disgust for humanity. It also gives a fresh twist to the old Gidean *"acte gratuit,"* which, for Sartrians, leads to what they call the necessity for choice, and has as its main premise a belief that there are two kinds of existence, one in the world and one in your mind, and as its slogan *"l'être et le néant"* ("being and nothing"). This belief leads to an anti-humanistic dualism and to a phenomenology which considers God a pure phenomenon—a notion thought up by the late Edmund Husserl, who has, along with other German philosophers, tremendously influenced the postwar French philosophers. These, now that Paul Valéry, with his theory of the apparent light of man's intelligence, is dead, regard Catholic Jacques Maritain as the last living example of French Christianism and have discarded deduction and rationalism and embraced mass responsibility and the doctrine that an autobiography of the thought of the thinker takes the place of thought.

Sartre, who did a remarkable series of articles as a roaming reporter in the United States last summer, used to teach in Paris at the Lycée Henri Quatre and today instructs his disciples mostly at the St.-Germain-des-Prés Café de Flore. In his Spanish War story called "The Wall," about a prisoner who was not shot, he has produced the finest short story of the resistance. He is also the author of a novel called, simply, "Nausea" and is already the best-known new Frenchman throughout Europe. Sartre is automatically fashionable now among those who once found Surrealism automatically fashionable.

Europe, warring and warred-upon for six years, is bankrupt but neglects to say so. Its editions of paper money say so for it. In Budapest the pengö is unofficially quoted at twenty-four thousand to the dollar. French peasants refuse gold sovereigns and American eagles. Gold, as a tangible value, has been out of style for so long that nobody knows what it is worth any more, or cares. Today peasants want bushels of paper money. It is a sign of the times. Industries want more money with which to make goods, premiers of governments want more money with which to govern, children prattle about thousand-franc or thousand-lire or thousand-mark notes. The war, which destroyed so much of everything, was also constructive, in a way. It established clearly the cold, and finally unhypocritical fact that the most important thing on earth to men today is money.

1946

January 3

On New Year's Day here this year there was more food on the plate, if less hope in the heart, than there was last year. A year ago there were still empty stomachs and heads were light not only with hunger but with desperate notions of a better world to come. The three most important things that have happened to France since last January 1st have been disillusion, devaluation, and the continued dominance of General Charles de Gaulle. All these happenings have been normal to the times. France today has little chance for the abnormal.

Achieving what amounted to a total surprise to the Paris population, the shops at Christmastime were suddenly filled with practically everything but the money to buy it with. Along with diamond bracelets, there were inelegant rabbit-skin moccasins—at the price of a midinette's weekly wage—for the icy, unheated home. There were grapefruit for the first time since 1939, at what a laboring man got then for four days' work. The butcher shops were crowded with the old-time, statuesque figures of sacrificial lambs, marble-white with fat and at such high, though legal, prices that the butchers have threatened to close up shop and hand the whole business back to the government, which has already taken back bread. Before the October elections, de Gaulle's government derationed bread, thus trading bread tickets for votes in a smart political move which elected its deputies, who are now faced once again with a wheat shortage. Under the new rationing, the government will give each person only three slices a day more than the Nazis gave in 1942, the worst of all years for those poorer French who practically live by bread alone. All over Europe, government has become a sort of harried, shabby, food-spotted maître d'hôtel fumbling over the menu of what its people will eat next and wondering who will pay the political and financial bill.

It is a pleasure to report not having seen the late Jean Giraudoux's last play, "La Folle de Chaillot," at the Athénée; its wildfire popularity, which makes seats almost impossible to buy, is an encouraging sign of revived public intelligence. The piece is already established as probably the most influential success of the modern French theatre. For several days after its première, the play was a topic for the editorial columns of nearly every newspaper in town. The theatre critics wrote of it with exaltation. In their various emotional and ideological flights, they compositely declared that one felt in the presence of a miracle; that it was a high French moment which the world envies France, the peak of Giralducien art, a lesson which we will never forget, the ghost of Beaumarchais, and an anticipation of the revolution which is coming. Provided the revolution does not come within the next fortnight and upset things, a few thousand of us who reserved our seats way back in '45 plan to go and see for ourselves.

January 30

The only flash of brightness that General de Gaulle's resignation brought was a political witticism which was not even true: "What we now have is a de Gaulle government without de Gaulle." Otherwise, his striding out of the Presidency left French morale in a state of empty gloom. This past fortnight has obviously been the worst here since the war ended. In Paris, the sense of crisis has been greater than in any other peacetime period since the February 6th riots of 1934, when there was at least the effervescence of fresh political hopes. The malaise is quieter but deeper now. There is no feverish surface reaction that might stir citizens into doing anything. The four years of occupation, the obedience to hundreds of orders, and the thousands of hours the French have spent in queues seem to have taken the old, violent, animated crowd spirit out of them. This is perhaps a good, perhaps a bad, thing; anyway, it is a new thing for Parisians. They act like people who know that there is nothing they can do except stand, as usual, and wait. The French are waiting to be governed.

Nobody, except possibly a few state librarians, believes that the General is really abandoning power in order to write his memoirs. A month before he resigned, the Bibliothèque Nationale, at his request, sent him a cartload of volumes by or on Mme. de Sévigné and the Duc de La Rochefoucauld, both of them certainly the wrong track for a well-educated military Frenchman who might want to brush up on the near coup d'état of General Boulanger (and his failure) or the successful coup d'état of President Louis Bonaparte (and his subsequent failure). Millions of French and the entire German Army found it difficult to believe that de Gaulle meant what he said when, in June, 1940, he declared that he would keep on fighting. It may well be that he was telling the truth again in January, 1946, when he said he would fight no more. De Gaulle is probably an entirely self-made democrat, having originally been an honorable Tory of the sort who would have done well in England a hundred years ago, and he must have decided, in recent months, that he could not give himself another complete education and become one-third Communist and one-third Socialist, like the Assembly he had to work with.

The editorials of respectful adieus to General de Gaulle were marked by a heroic, elegiac tone, as if it were still he and not his critics who set the style. On the Left, his captious Socialist admirer Léon Blum said that his departure was "dolorous" and "perilous for France, since it created a void equal to his services and the place he had held in the nation's life." Over toward the Right appeared statements and phrases such as "He retired with his dignity intact. He remains one of the high summits of our history . . . his romantic heroism . . . his ascetic scruples . . . his lack of Machiavellism." The General is still France's unknown soldier. No two people who have worked with him, and no two who have analyzed him from afar, seem to have come to the same conclusion. He has been earnestly described as a male Joan of Arc, as a Hapsburg type born too far north, as an inspiring, almost royal figure, and as the last great Republican legitimist. It has also been said that his blazing sincerity and lack of adaptability were against him. In his favor it has been admitted that he has an amazing sense of timing, his only political gift. Perhaps the cruelest and truest comment is that his leaving now is the greatest service he could render to France. In a tragic, ironical way, he symbolizes her shame, her defeat in the war. Now that he is off the scene, France can move on and try to forget.

It isn't often that the clear-cut success of a play here—or anywhere else, for that matter—can be estimated by the amount of confusion it arouses. It is a real tribute to say that this is the case with Jean Giraudoux's "La Folle de Chaillot," which means "The Madwoman of Chaillot," to begin with. People are already addled, on their way in to see it at the Athénée, because they have heard far too much about its snob, revolutionary, and intelligentsia importance. They are completely muddled, on their way out, because it is impossible for them to believe that they've just been through all they have and that it gave them such a delightful time. The plot goes as follows: Aurélie, an elderly female from the Chaillot section of Paris, who enjoys the privilege of sitting in her tatterdemalion finery on the terrace of the Chez Francis Café, there overhears a goldbrick promoter planning to fleece the public once more by selling it shares in a petroleum company which, this time, will claim to have struck oil in the very foundation of Paris—its sewers. Calling to her aid her friends, who include, among others, a waiter, a hat-check girl, a ragpicker, a sidewalk singer and juggler, a deaf-mute beggar, and a half-drowned young man who has been rescued from an attempted suicide for love, she plans to save society by eliminating the stock market. That is Act I. Act II (there is no Act III) consists of the elimination. After taking counsel with three other lady lunatics, from three other sections of Paris—one lunatic is accompanied by an invisible dog she loves—and after ascertaining from a genial city-waterworks employee that the eerie basement room she lives in has a secret door which leads down to the sewers, Aurélie decides to give a party. She invites the petroleum-company promoters, their high-society lady friends, their press agents, and their government lobbyists to visit her and their subterranean oil fields, and, after the last businesslike crook and pal have disappeared underground, she simply locks her secret door. That is all. Or anyway, all there is to the much-discussed revolutionary implication of the play.

The love theme is fancier. It involves Aurélie's thinking that the young near-suicide is her own youthful, untrue love, Adolphe, come back to betray her again, which delusion, in a dream sequence, gives the play a sort of mid-last-act happy ending. It is just after the petroleum promoters have disappeared underground to die that up the sewer staircase march three dead Adolphes in attractive cerements—a sort of Seducers Corporation, perhaps, since they state that they represent all the perfidious Adolphes ever on earth. The play ends with the café-terrace mountebank crew swarming in to take the four lady lunatics off to a really pleasant party, and with Aurélie's philosophic comment that "these are the people who are actually worth while." It might be added that the promoters, society ladies, and lobbyists, on their death march, are followed by two other strange companies, listed on the program as The Friends of Vegetables and The Friends of Animals. Perhaps their role was to attract the gnats that the audience strained at and the camels it swallowed.

What makes this play more remarkable than even the plot may lead one to suppose is the work of four greatly gifted people involved in it. Its production is a brilliant tour de force by Louis Jouvet, possibly the most creative actor-manager in Europe today. He plays a minor part, that of the ragpicker. There are fifty-seven characters in "La Folle de Chaillot" but only one real role, that of Aurélie, a nearly endless monologue, recited with a lifetime of technique by the elderly, energetic Marguerite Moreno. She dates from near the founding of the Third Republic and is a notable surviving example of that early vintage of gifted Frenchwomen in which her friend Mme. Colette was the literary champagne. In a program note on Moreno, Colette writes, "Her name has often figured in my books, like a clover leaf pressed between the pages." As the old lunatic, Moreno wears a startling makeup—the white face of a clown, with the wise, black, saucer eyes of an owl. Her foolish finery—such as her great hat, with its mad white bird, her sumptuous, silly dress of red-and-green silk, with torn black lace drooping at half-mast—and the astonishingly beautiful basement room are the careful, *démodé* romantic creations of the painter Christian Bérard. His old-fashioned, elegant white dress for the dog-loving lunatic is like a page of embroidery from Marcel Proust. One of the other lunatics is played by Lucienne Bogaert, now almost unrecognizable, like many once beautiful sights in Paris.

The fourth artist concerned is, of course, the author, the late Jean Giraudoux. American theatregoers who recall the charming, anachronistic titillation of his "Amphitryon 38" might be less surprised at the fantasy of his last piece than at the heavy, symbolic political significance Paris gives it. Maybe Paris is right. Just before his death, in 1944, Giraudoux prophesied, correctly, that his play could be appropriately produced late in 1945. Thus his bitter whimsy about a righteous madwoman suppressing a wicked big-business oil company runs full tilt into the beginning of 1946, just when the country's big-money gas and electric companies, following the banks and mines, are being scheduled for nationalization; just when the arrest of the big-money gentleman who owns the Paris Métro has been demanded by the Left press; and just when thirty-one of the big-money French insurance companies have been jerked up against the law for having insured the big-money black-market food racketeers against any possibility of loss—a pretty whimsical idea in itself. Giraudoux's play is no Beaumarchais "Barber of Seville," in respect to revolution. But it is an oddly fashionable sign of the tensely speeded up French evolution from Right to Left, from what remains of the bourgeoisie to what is coming from *le peuple.*

May 22

During the three weeks in which the Big Four failed to get on with the peace, Paris took a big step forward toward what must be one base for it—prosperity. The intentions of our quadripartite statesmen toward the Senusi tribes in Cyrenaica are still obscure, but anyhow the Galeries Lafayette and seven hundred and ninety-nine other Paris shops have just placed on sale the first postwar men's, women's, and children's shirts, pajamas, and underwear. New underdrawers are vital to a civilized reconstruction of Europe, and as a part of the peace which diplomats had nothing to do with, they seemed wonderful. The garments were sold one to a customer, under the category *"Articles Utilitaires,"* were priced cheaply from a hundred and twenty-five to two hundred and fifty francs (a franc is now worth four-fifths of a cent), and took only one textile ticket off the family ration card. There were not nearly enough to go around. Under the planned-economy program of the Ministry of Industrial Production, they were made up by the shops themselves from cotton woven only this year in the north, which is France's spinning center, and every garment looked like an advertisement for the Communist Party. On each was stamped, along with the French cock, the phrase *"Renaissance Française"*—the French Communists' postwar slogan. It was as though the New Deal, during America's depression, had sold drawers marked "For the Forgotten Man."

The inexpensive and unexpected underclothes are only part of the astonishing revival in Paris, which, starting tentatively with New Year's gifts, has snowballed modestly. Until recently, Parisians and their show windows were both shabby. Now the windows, at least, are excellently dressed. Articles that spring from good taste and that look luxurious, homely necessities, and small comforts, all of which had become merely memories to Parisians because of the lean war years, are now suddenly pushing back into circulation and, taken together, give a surface impression that Paris is struggling back to what looks and feels something like subnormal, at least. In other words, Paris is now like a thin, ill, handsome old woman with some natural color flushing her cheeks as she tumbles to her feet. Paris is still possessed of her remarkable habit of survival. She has already received two great boosts from outsiders, pulling in opposite directions—the recent American loan and the earlier impulse toward disciplined work which came from Russia. The strictly European countries cannot aid one another. All of Europe, France included, is sick, weak, and physically and morally hungry in varying degrees. It is precisely in her stomach that Paris has recovered least. Parisians can now lunch off a legal menu in a very modest restaurant for three hundred francs, if they can afford it. The white-collar and the working classes cannot.

As if seeing new historical sights, one's eye instinctively makes its list of cheering novelties as one walks the streets. For one thing, there are almost twice as many newborn babies on view this year as last. These new citizens can be seen all over town, bundled up for protection against drafts in the Métro, spread out on their mothers' knees to the faint sun in the Tuileries Gardens, or rumbling along the sidewalks in what is left of French baby carriages, which, as transport vehicles for everything except babies, took a terrible beating during the war. It has been announced with satisfaction that one of the Ministry of Industrial Production's next offerings of utility articles will feature inexpensive baby carriages.

Transportation for adults has already expanded. There are now five thousand taxis in Paris, and they are available even for healthy or unimportant people. Last month, you had to be pregnant, ill, or on top-level official business to get a taxi, and you had to apply first to the nearest police station. These liberated taxis are too few to accommodate all, but at least they officially are cheap. It costs about twenty-five francs on the clock to drive from the Opéra to St.-

Germain-des-Prés. The catch is the *pourboire*. The proper tip would be around fifteen francs. The Invalides aviation station for plane passengers bound to and from the suburban flying fields is being hustled to completion; it was started only three months ago. Twenty-seven more of the closed Métro stations have just been reopened, which means a saving of miles and hours of walking. Some of the antique green autobuses have lumbered back on the job, all with new routes and numbers, thus mixing confusion with satisfaction. Now that Parisians are off their feet more than they have been for six years, a ration ticket for having shoes resoled is to be distributed. Five-horsepower Simcas, France's midget motorcars, are now rolling out of the company's Nanterre plant at the rate of forty a day and can be seen passing through the streets en route to Belgium, Switzerland, Holland, and Denmark. The French will not be allowed to buy them until later, and then the price will be 92,000 francs. The new eleven-horsepower Citroëns will cost 121,180 francs, precisely. France has tripled her exports in the past four months, which may be encouraging reading for Americans, but the inflated banknotes that have been appearing in steady volume are gloomy reading for the French. Despite the upward swing of business, another new banknote is just out—a fancy, five-hundred-franc note stamped with a portrait of the melancholy poet Chateaubriand holding his head, as well he might.

Fifty-three Paris theatres and five music halls are going full blast—a sign of the continued restless tension, the hunger for unreality, and, indeed, the plain hunger that still obtains. It is cheaper for a modest bourgeois couple to nourish their imaginations at a theatre than their stomachs at a black-market restaurant. The Folies-Bergère has just put on an eye-popping new show, with the best-dressed girls and the best-looking undressed girls that it has assembled in ages.

The new comedy hit is Marcel Achard's "Auprès de Ma Blonde," at the Théâtre Michodière, starring Yvonne Printemps and her husband, Pierre Fresnay, in five stages of contented marital love. This is absolutely satisfying second-rate theatre animated by an enormous cast, all playing perfectly, as if with first-rate material. The chronology, which runs in reverse, starts in 1939 with rich old M. and Mme. Chose happily celebrating their golden wedding anniversary, and from then on goes back to 1889, when they ran away together, penniless, to get married. The disagreeable children they subsequently produced provide the rest of the plot. The play has three virtues and one error. The error is that Mlle. Printemps' role never permits her to sing. The first virtue is that the play allows her to start the evening as a gray-haired, deaf, but dexterous old *grande dame* and then romp back into turbulent, blond youth. The second virtue is that four of the acts take place in the same salon, which marches backward in its décor from prewar chromium-steel modern to 1899 Lalique, the room and its changing appearance thus acting like a living personality. The third virtue is that the ladies' costumes were created by Mme. Jeanne Lanvin, *la grande couturière*. With needles and braid, she has written a sort of brief, feminine French history on satin and wool.

June 12

When the Opéra put on a gala ballet night last month for the Big Four conferences here, it hung a sign over the box office timidly requesting the public to wear evening clothes "to the extent to which it may be possible." The invitations for the Galerie Charpentier's recent evening première of its modern-art show, "Cent Chefs-d'Oeuvre des Peintres de l'Ecole de Paris," peremptorily told those invited that dressing up was "acutely desirable." This dressy upper crust was asked free. So many of the lower crust were eager to pay the five hundred francs' admission in order to be seen in the upper crust's company—and probably in order to see the pictures, too—that a riot squad was called to handle the crowd that choked the Faubourg St.-Honoré, and half the élite, including Cabinet Ministers, were unable to squeeze in among the art.

Another controversial art exhibition has opened at the Orangerie. It is the first big postwar government show and bears the long, explicative title "Chefs-d'Oeuvre des Collections Françaises Retrouvés en Allemagne par la Commission de Récupération Artistique et les Services Alliés." The Allied Services referred to are the United States Army's Monuments, Fine Arts, and Archives Sub-Commission, and the art recuperated from Germany was looted during the Occupation from the houses of famous Jewish collectors in Paris and for the most part handed over to those noted connoisseurs, Hitler and Göring. Since these pictures came from private collections, the public has been interested in the first chance it has had to see gems like Vermeer's "Astronomer," which went to the Führer; Fragonard's "Girl with China Doll," which Göring held out on Hitler; van Gogh's "Bridge at Arles," which Göring gave to Frau Emmy; and Lancret's "Le Repas de Chasse," which bears a bullet hole it got when it was shot up in the Reichsmarschall's loot train during the battle for Berchtesgaden. Privileged citizens who used to admire these pictures when dining *en famille* with the Rothschilds or the other noted collector-owners were doubtless even more interested to learn from the catalogue that "Creation of the World," which has always been ascribed to Patinir, and "Adam and Eve," which generations of students have been told was by Cranach, are not the real thing but merely belong to the respective schools of those painters. Out of three portraits which have always been considered Bronzinos, two are now merely attributed to him. A picture long catalogued as an elder Breughel, an "Infanta" that people thought was by Velásquez, and a portrait of the little Princess Henrietta Maria of France for which Van Dyck has been given credit for centuries are in a similar fix. The exhibition has developed political aspects, too—the world being what it is today and what it was yesterday—because the catalogue fails to give the owners' names. This omission is being interpreted as, first, a sop to fear of Communism, since the owners are rich, and, second, as a sop to fear of a revived Fascism, since the owners are also Jews.

The French state proudly takes the view that these looted pictures are a part of the patrimony of France herself and has insisted on bearing the very considerable cost of assembling, packing, and restoring them to their owners' hands in Paris. Some liberal art circles here think that a third sop (one to democracy) is now indicated—that out of thankfulness to the French Republic for its respect for religion and private ownership, gifts should be made to the state's public museums of some of these invaluable, authentic masterpieces, which the taxpayer could then enjoy on any Sunday afternoon. Léon Blum's wisest postwar aphorism is that the prewar stupidity and selfishness of the bourgeoisie and of the rich cost them the political leadership of France. Their now giving a portion of their possessions to the state and of their brains to its problems might save them from having to give their heads and all the rest later on.

June 26

In June seasons before the war, one of the big French racing fixtures was called the President of the Republic Race. When it was run this year it was still called that, but it should have been renamed, to be politically up to date, the President of the Provisional Government and Minister of Foreign Affairs Race, these being the titles under which the newly elected Georges Bidault functions as chief of the French state, following the latest Communist-Socialist M.R.P. (Popular Republican Movement) upheaval. This tripartism had a definite effect on most of the activities that went on during what developed into a hoity-toity June season—the first in seven years. Most of the social events were class-conscious and charitable, and the finger of the French Red Cross was in nearly every pie. The Red Cross got off to a bad start with an announcement of its prices for a concert to be given by Toscanini and the Milan Orchestra— eight thousand francs a seat (a stenographer's monthly wages), which people indignantly called an *insulte à la misère,* or plain gouging. It probably did better with its Pré-Catalan gala evening in the Bois, for which the ladies were requested to wear masks. Thus a few square inches of black satin on the face substituted for the fabulous costumes that used to swathe the demoiselles from head to foot in the famous prewar *bals masqués.* But the Red Cross failed with its Nuits Foraines, a three-night gala at the Théâtre des Champs-Elysées, the parterre of which was boarded over so that a street fair could be installed. For one reason and another, the Nuits Foraines came in for considerable criticism. Memories are long here. Among the war-aid organizations that worked in France during the Occupation, the American Quakers and the French Salvation Army are more popular in retrospect than the French Red Cross.

The most indestructible talent in Paris is still that of Jean Cocteau, whose new ballet, "La Mort de l'Homme," has just been given its première by the Roland Petit Ballet, at the Champs-Elysées Theatre. It was one of the high moments of the season. The audience felt the same sense of astonishment that Cocteau aroused with everything he laid his pen, pencil, or imagination to between the two World Wars. The passage of time seems neither to wither nor even to interrupt the hothouse ripeness of his talent. What his startling new romantic ballet contains is this: A youth in his mansard room awaits his beloved, who arrives and refuses him, pointing to a gibbet that suddenly becomes visible beside his garret window. He hangs himself. At this moment the garret walls disappear, to be replaced by the nocturnal Paris sky, lighted and dated by the Tour Eiffel and its prewar Citroën sign. Then Death enters, wearing a gaunt mask, which, when lifted, reveals that Death is the beloved girl. She places the mask on the youth's face and gently dances him away from human view. Finis. The ballet was hailed as a triumph, even for the prodigious Cocteau. Characteristically, he rehearsed it to jazz, to obtain the rhythm he wanted, but had it danced to Bach. As the moribund youth, Jean Babilée, a popular, handsome new French dancer, leaped and died with perfection.

It would be ungrateful to deny that a regime of civilized, almost willful perfect taste and artistic creation was maintained in prewar Paris by what was known as *Tout Paris,* which consisted of perhaps a hundred individuals out of a population of two million. This special, experienced crew were the patrons of fine clothes, just as they were the patrons of fine arts, and they collected early Chanels precisely the way they collected early Picassos. It has seemed fair enough to wonder, rather morbidly, whether this dominant small set of internationally overpublicized Parisians would still have the strength to manipulate the selective sense of this city after a war that brought greater changes than any which their immediate privileged ancestors survived. This year's June season has given the answer. Those who once succeeded in making their particular kind of season go have finally started to fade from the picture, and their particular kind of season is fading out with them. They are finished. They are the ones who are now démodé. For a long time they helped keep Paris the so-called civilizing capital of Europe and did a whale of a lot for the dressmaking trade. This trade, with its multiple beautifying ramifications, was the second largest exporting industry in France and gave employment to two million workers. The chic in an evening gown today costs twenty-five thousand to forty thousand francs. Some of the elegant women who helped set the style in the June season of 1939 are, in 1946, wearing the same '39 gowns. Only the black-market *nouveaux riches* can afford to dress all new all over. The best report on the Paris of today was written by Balzac, in his "Scènes de la Vie Parisienne." There has been a season, true enough, this year, and it has been an active one. The race tracks, the opera, the good theatres, the costly bars, and the hotels have all been packed. But it is a new Paris and a new public, both now only two years old. And temporal power, and all that goes with it, has completely changed hands too. The latest Cabinet of Ministers under whom France is living comprises six former professors, six lawyers, three journalists, three metal workers, three white-collar workers, two miners, two electricians, one engineer, one pharmacist, one broker, and one lonely rich industrialist.

In the midst of the new, Paris is also certainly leaning heavily on its past. The result is not a mixture but a blend. The greatest exposition of French tapestries—which is at the same time one of the greatest art expositions—ever held here has crowds of citizens queuing up at the Musée d'Art Moderne, on the Quai du New York. The show begins with the famous fourteenth-century "Saint John the Evangelist" tapestries, formerly hung on gala days in the Angers Cathedral, and ends with tapestries patterned on Lurçat's "Modern Roosters," which wear leaves instead of feathers. Two women have been influential in the making of tapestries based on modern French paintings. One is Miss Alice B. Toklas, Miss Gertrude Stein's friend, whose early grospoint embroidery of footstools based on designs made for her by Picasso showed new possibilities in tapestry. The other is Mme. Marie Cuttoli, wife of the senior senator from Algiers. Her early sponsorship of tapestries based on paintings by Braque, Dufy, Matisse, Léger, Le Corbusier, Miró, Rouault, and Lurçat has resulted in some of the most interesting tapestries in today's show, which includes everything through the rich years of French weaving—the blue-backed *milles fleurs,* the Gothics (with their unicorns), the Aubussons, and the fat nude and hatted females of the early nineteen-hundreds. Lurçat has taken up tapestry-making himself and the show also includes a reproduction of his workroom—with his designs, graphs, looms, threads, and wools—and exhibits of what Gobelin and other manufacturers are doing to revive one of the great, agreeable industries of France.

July 25

Paris has certainly been animated. Crowds have gathered on the boulevards in the past week to shout, "Death to

Daladier!," which they maybe didn't mean, and "We want our twenty-five per cent!," or the big general wage increase, which they are very serious about. The government is now asserting that it will authorize only an eighteen-per-cent boost, though it is unable to deny that the cost of living, dying, and being ill has certainly gone up forty per cent in the past year. The latest figures give eighteen thousand francs a month as the average H. C. of L. for a Paris couple with two children. Schoolteachers, who last month held a parade protesting their salaries, earn nine thousand francs a month and cannot afford any children other than those in their classes. Inflation is covering Europe like a magnifying glass. Seen through it, objects swell to twice their size in necessity or desirability and the piles of paper money grow to distorted heights. The inflation in the United States is regarded as even more grave. The dollar has fallen on the black market here from three hundred to a hundred and ninety francs. At this rate, it soon will not pay expatriate Americans to be dishonest.

The recent meetings of the Assembly were the most violent that the Fourth Republic has yet weathered. They were important because the Assembly majority refused to unseat deputies the people had elected. They were important because, for the first time in a crisis, there was no babble about France's past glories. There was, instead, both inside the Assembly and among the men on the street, a unifying feeling of war guilt—the guilt of brainy, weak statesmen, the guilt of greedy, middle-class businessmen, the guilt of partisan workers, all of whom, a family of forty million, had let their house fall.

August 1

A throwback to the Treaty of Versailles days was furnished by two of General de Gaulle's recent speeches, the first of which was given against the carefully selected backdrop of the Vendéen countryside, from which Clemenceau came, and the second at Bar-le-Duc, which was Poincaré's country. The first speech was on France's *politique intérieure,* the second on her *politique extérieure,* topics which let the General cover a lot of ground. What he said in his second speech not only was offered in the talented, pungent literary style that is characteristic of all the General's versions of French history but contained classic evaluations of the new world balance of power, evaluations which make some of the Peace Conference practices look like a reckless game of teeter-totter, with the twenty-one big and little boys trying to outweigh each other at opposite ends of the board. "Lucidity and strength of soul are what we have need of," de Gaulle boldly began, "in order to consider frankly, and realistically, the situation of the world and the position of France. The peace of France depends primarily upon the determining of the destiny of Germany. As a result of France's wounds, the equilibrium of the world has been compromised. Whatever Germany has been through, she remains Germany still—a large nation, a mass of people living in the heart of Europe—who, though she is in the depths today, remembers her days on the heights and whom the daemon of war might one day tempt again if she has the chance to regain her greatness by mating her ambition with that of someone else. She must be permitted neither to tempt nor to be tempted. If not, woe to the sons and daughters of mankind! Nobody in the United Nations should have in the back of his mind the idea of using the renaissance of Germany as a threat against another nation. As a result of the thirty years' war we have just been through [as a military man, de Gaulle always considers that the armed peace of 1918–39 was merely an interim diplomatic battle between the first and second sessions of one World War], a cyclone has passed over the face of the world, and with the collapse of Germany and Japan and the weakening of Europe, Soviet Russia and the United States of America today are alone in the front rank. Rich in their men and their resources, which are all within their territories and naturally protected, in the case of one by immense oceans, in the case of the other by its own space, both are drawn toward expansion, which, according to eternal custom, is clothed in the mantle of doctrine but is in the end an unfolding of power. The emergence of these two new world powers coincides with the discovery of terribly powerful methods of destruction. The United States of America's just and good proposal to place fissional activity under international control is a duty toward humanity that overrides the interests and claims of every regime and nation. And if this duty be not fulfilled, clouds of danger will cast their shadow over every living thing."

To Gertrude Stein's old friends here, her death was the last chapter in her private history's concordance with the important things going on in France. She had lived in France for forty years, had worked for it during its two greatest wars, and had received public acclaim here during the period between them; she had met and welcomed our Army during the liberation; and she left the scene only as international statesmen began talking over the second peace. All the French newspapers, in their obituaries, mentioned the friendship she and her faithful companion, Miss Alice B. Toklas, had for Picasso and Matisse. As a matter of fact, it was the ladies' unchanging good relations with modern paintings rather than with modern painters that best demonstrated the solidity of their jointly operating critical faculties. As they frequently choroused, painters were to be admired for their paintings, not for their characters; first-class painters often had difficult characters, and only third-class painters had really good characters. Most American collectors eventually bought modern French art as they might have bought cut flowers from a florist's shop. Miss Stein pulled hers fresh from the stem, in the ateliers where they grew, four decades ago. Today, her collection is as remarkable for its ripeness as for its freshness. Miss Toklas says that she has no idea what Miss Stein wanted done with her pictures; she says Miss Stein never made any plans for her art, she just enjoyed it.

August 21

Paris now belongs to the peacemakers. It may not be the kind of Paris some of the Conference delegates expected or want. Mostly, its theatres are dark, its restaurants locked up, its little shops closed. Three-quarters of a million Parisians, or over a quarter of the population, finally succeeded, with difficulty, in getting themselves and their luggage out of town by Assumption Day, and wherever they managed to go, they will stay until August is out. There hasn't been such an exodus since the tragic one of 1940. What with the worry about Munich in 1938 and the fear of war in 1939, most people here have not drawn a carefree breath in eight years. Nobody has yet forgotten what happened during those years, but this month there has been at least a readiness for a change of scene. This is the first big postwar holiday. All classes have participated. On the front doors of little shops are pasted bits of paper, in cramped script, advising *"Notre Gentille Clientèle"* that, at the request of the employees, the shops will be shut until September. On Thomas

Cook's door is a large sign warning de-luxe travellers that accommodations in *wagons-lits* are no longer available except a month in advance. Taxi drivers have been asking you if you wanted to ride down to the Riviera with them and their wives for four thousand francs per passenger. Only the French can afford to go there, however. Paris is dear, the seashore is dearer, but for once the franc is strong. It is the legal-rate dollar and pound in this year's tourist's pocket that are in the doldrums. British pleasure travellers are allowed seventy-five pounds a year to spend outside their island, and that has not carried the British far in their brief tourist invasion here. To Americans, this is the memorable summer when Paris shopkeepers can say primly, for the first time in this century, "It may be expensive for you in dollars, Madame, but it's not expensive for us in francs." It is the summer of the Peace Conference, of the sound franc, of the slowly dying black market, of the reviving supply of goods.

The Congress of Vienna, which gave Europe its longest recent peace, concerned itself with the balance of power in Europe. If anything is clear about this present Paris Conference, it is that the balance now being sought is a balance of the world. At the Congress of Vienna, all the winning parties saw eye to eye, except, of course, as to how much each ought to get; they believed in the same over-all fundamentals—in kings, Christianity, the supreme importance of Europe, and a settled, three-class society of the rich, the middling, and the poor. It is the difference in what Byrnes and Molotov, taking them as chiefs and symbols, believe in that makes this Peace Conference a battleground of opposing ideologies.

One can report with pleasure, the Louvre has been holding a small and exquisite de Goncourt exposition which offers a review of a literary age of logic, sensibility, and comprehensible gossip, and which commemorates a coterie of modern geniuses who are still the chief ornament of today's French letters. There are portraits, photographs, private letters, pink waistcoats, walking sticks, first editions, cartoons, and even the tables and chairs that belonged to men and women like de Maupassant, Zola, Coppée, Flaubert, Huysmans, Dumas, Hugo, George Sand, Heredia, and Renan. It also tells, beneath Flaubert's portrait, what the de Goncourts said of him: "Tall, with big shoulders, with big, beautiful eyes popping out of slightly swollen lids, drooping mustache, and a mottled complexion specked with red." De Maupassant also went into details about Edmond de Goncourt: "He has long, grayish hair, which looks discolored, a slightly whiter mustache, and singular eyes, spread out by strangely dilated pupils," as indeed they may well have been, busy staring with astonishment at his great generation, living for talent, for the perfected word, for the fight for civilized ideas, in a Paris of top hats, bustles, peace, candlelight, and time to think.

October 1

In the corridors, there is a strong feeling that the reason the Peace Conference conclusions appear abnormal, spongy, and unstable is that the new balance of power in Europe is now being balanced by a couple of powers that are not European—Russia and the United States. For once, here is something that is not the peace delegates' fault. The situation is, axiomatically, part and parcel of the strange, unidealistic, second twentieth-century peace which this last war, essentially one of gigantism in manpower and production, has loaded onto the U.S.S.R. and the U.S.A., the only two states big enough to take it. Because of sheer size, they have become the world's bosses. They have set up the chimera of absenteeism as the new order for Europeans, some of whom may be the most civilized creatures on earth but all of whom are now reduced to sitting quiet and worried in their cities or on their land and nodding like yes men to a pair of enormous, newly grown-up outsiders. Europe has lost the right to maintain her own power balance because no sooner had it been reëstablished by peace, after one war, than uncontrolled Germany upset it and the world by rushing us all into battle again. Nevertheless, to anyone familiar with Europe (including England) in the thirties, which were doubtless one of Europe's rottenest periods, owing to the reckless policies pursued by the main democracies (policies in which, as proof of our coming responsibilities, we distant Americans shared), there is something sickly in the spectacle of Europe's becoming a conglomerate, largely literate colony, eyed by Russia, which cares, and indifferently tended by the United States, which, except commercially, really is not interested. After two thousand years of Europe's dominating civilization, a nervous shiver runs down the spine of any thoughtful American at the Peace Conference when he realizes that the twenty-one nations' voices are actually silent, that the Big Four's voices are not even a strong and constant quartet, and that the Big Three's voices are not always a powerful trio, because what counts is only the voices of the Big Two, the voices of the East and the West. The world seems to have lost two points on its compass. The earth's surface is changing. Europe is contracting; the U.S.S.R. and U.S.A. areas of influence are expanding. It is as if Europe were slowly entering a new ice age.

1947

January 2

France's Christmas gift to herself was her Fourth Republic, generously presented a few hours before the dawn of December 25th. In making this gigantic nocturnal offering, the presiding chairman of the Council, as the Chamber is now called, said to the weary councillors, "I salute the birth of the Fourth Republic. My best wishes to it are that its new constitution, adopted by the people, may affirm our Republic and our democracy. Long live the Republic! Long live Liberty!" There was applause. The Fourth Republic, like the Third, is a result of a defeat by the Germans. (In 1870, a revolutionary Paris Commune was in the offing, a prophetic counterpart of the Communist Party that functions today in the Parliament in elected and orderly quasidominance.) France's newborn Third Republic was so rich that it immediately paid its war debt to the enemy Prussians and could have paid it twice; the Fourth Republic starts so poor that it has already borrowed from a transatlantic friend. The Third Republic was founded on the curious notion that it would conveniently collapse as soon as the next Bourbon could be maneuvered up to the throne. The Fourth Republic has in part been set up to shut out General de Gaulle, the only massive figure still visible, though faintly, on the French landscape. As the newest European republic makes its difficult first motions of life in a France whose violent idealism created Europe's first modern republic, the belated suspicion is stirring among French citizens that perhaps republics, though conceived strictly on paper, take after each other, in hereditary fashion, much as kings do, and that by some law of genetics the Fourth Republic, when it is a few years old, may begin looking and acting like its parent, the so-called harridan Third Republic, which was no beauty.

In one way, certainly, the two republics will differ. The late No. 3 was so infested with political parties—anywhere from a half-dozen major groups to three dozen or more minor schismatic rivals—that government, enfevered by the multitudinous political bacilli, was constantly dropping weakly in its tracks. In No. 4, the parties are few and hardly count anyway. What does count is its two grim, directly opposite political directions, Left and Right, or Communist and M.R.P., with literally equal strength, so this time the government may not fall, simply because it may not even be able to get going. The Fourth Republic, by a triumph of balloting and the democratic process, can sit perfectly paralyzed by what temporary Premier Léon Blum has described as "those two great parties whose simultaneous presence in government is at once indispensable and impossible." A sardonic Deputy, discussing the semi-existence of the two minority parties, the Socialists and the Radicals, whose thin shadows of strength will be thrown, respectively, to the Left and the Right in efforts to create majorities, declared that in a full-length anatomical portrait of the new government the head and both arms would belong to the M.R.P., the neck, heart, lungs, and left foot forward would belong to the Communists, the Socialists would have the stomach and left leg, the minor Left Wingers would be punished by having the right leg, and the reactionary, pro-Church P.R.L. (Parti Républicain de la Liberté) and its hangers-on would have the right foot, including the Pope's great toe. The First Republic, with the great words "Liberté, Egalité, Fraternité," produced a portrait of man's invisible regions. In the Fourth Republic, what the French have retained of this sacred trio is a too exact equality between Left and Right. Some impulse, whether shock, force, or weakness, will most probably cause France to move, eventually, in one direction or the other.

The Paris theatre scene has been enlivened by the non-commonplace—not necessarily successful as entertainment but distinguished because it comes from important pens. One item is André Gide's translation of "Hamlet," playing to dutiful, admiring crowds at the Théâtre Marigny. Because Gide now ranks as a demigod, no one has had the pluck to say that his translation of the poetic masterpiece not only lacks wings but is downright flat-footed modern prose. For "Get thee to a nunnery," Gide substitutes what sounds like an address tossed to a taxidriver—"*Au couvent.*" For the tragedy's great epitaph line, "Good night, sweet prince," he offers merely "*Bonne nuit, gentil prince.*" (According to Cassell's bilingual dictionary, "*un gentil enfant*" means "a nice, amiable child.") Since, because Gide's name is attached to it, the play's the thing at the Marigny, little attention has been paid to the production, which is one of those athletic, hearty, hasty "Hamlets," with everyone but the ghost going on- and offstage on the run. Jean-Louis Barrault, as the nice, amiable prince, is remarkable in the death scene, to which he gives an oddly cinematographic effectiveness.

Another special item has been the Théâtre Antoine's program of two lengthy playlets by the Existentialist Jean-Paul Sartre. Both deal with torture, the first in France and the second in our Deep South. The opening piece, "Les Morts sans Sépulture" ("The Unburied Dead"), shows two sides of the French Resistance: first, the patriots' side, through a group of captives in jail, including the inevitable pretty girl, and a philosophic analysis of their arrogant human pride in enduring agony without screaming, let alone squealing on comrades still at liberty; second, the captors' side, in the room next door, where Vichy Frenchmen do the torturing right in front of the audience's startled eyes. In the Vichy group is another inevitable character, the neat, blond sadist who loves his work and the music of "Tosca." He sees to it that the patriots, after they have argued themselves into being spared, are shot anyway. This sort of masochistic entertainment, once idly popular at the Grand Guignol, has now been reclassified as part of recent French history. Sartre's second playlet, "La Putain Respectueuse" ("The Respectful Prostitute"), must be the world's only funny play about a Deep South lynching of a Negro. It is a brilliant, disturbing mixture of melodrama, farce, and deadpan reporting on the Bilbo mentality. The trollop of the title, who at first tries to save the Negro, falsely accused of having accosted her, furnishes the only truthful, and thus the only moral, element, which is finally nullified because of her eager, noisy desire to be recognized socially by the family of the local senator—the uncle of the leading lyncher—at whose request she finally bears false witness against the Negro. This unhappy victim is played, with poetic postures of fright surely never seen in our Cotton Belt, by Habib Benglia, an ebony mime formerly famous at the Folies-Bergère. In addition to the extraordinarily lifelike, funny, kind, ebullient, coarse-mouthed, rattlebrained figure of the whore, there are characterizations of American male prudery, false patriotism, hypocrisy, sentimentality, and what looks to foreign eyes like all-around national infantilism. These are the spicy, comical ingredients in Sartre's deadly skit, the main weakness of which is that it wavers between being a real satire on the Southern U.S.A. and a sort of shapeless exaggeration of it. As a consequence, the play has an amateur quality. This quality may also be due to Sartre's having met so few authentic Southern-gentleman lynchers on his recent American lecture tour.

France's second New Year of the peace opens with an Eastern war. The new Republic's preamble to its constitution states that it will "undertake no war with a view to conquest and will never employ force against the liberty of any people." That is precisely what it appears to be doing in Indo-China.

January 16

Léon Blum's now completed thirty-one days as France's pro-tem chief are, like his prewar Popular Front period, peculiar, very personal islands on the map of modern French history. As proof that he is, perhaps fortunately, a poor politician, both times he was on intimate terms with the French populace, ordinarily in contact with their politicians only through the ballot. Blum, a bookish intellectual who wandered into politics straight out of his library, has social aims that are humanistic rather than revolutionary. His recent governmental announcements to the nation read like confidential correspondence with a group of friends. Almost every other night for four weeks, he carried on what sounded like one side of a long, interesting conversation with France. Before New Year's, when he began begging the French to help him save the franc by being unselfish, it was the quality of his mind and hopes, against the background of all his years of public life, that—after the first moment's surprise—animated a good will that spread from house to neighborhood, from factory (after a few misgivings) to shop, from farm to suburb, and pretty well all over France. The famous five-percent price cut that he offered was no dry financial program but a typical, even cozy, proposition of friendliness. Everybody expected that this year's living costs would jump a mile high, which is just where Blum himself jumped them to balance the budget, but then he also brought them down an inch, which was what no one had expected.

Of all France's old, prewar figures who have again become active, Blum alone has really come full circle. He was a young Jewish littérateur in Paris during the Dreyfus affair, which first divided modern France into Left and Right and into Catholic and anti-clerical. During the period of his Popular Front, which marked the triumphant emergence of the Left, his allies the Communists, then also enjoying their first important party victory, refused to accept any Cabinet appointments or participate in power. It was during that period that his name began to serve as the nobler half of the premature collaborationists' slogan "Better Hitler than the Jew Blum." While Hitler was still having the best of it, the Nazis sent Blum to Buchenwald for his triple crime of being French patriot, Socialist, and non-Aryan. Blum has now made his final, elderly reëntry into, and, it appears, his weary, promised exit from, politics at the exact moment when his former allies the Communists not only are the leading Communist party of Western Europe—and by a narrow margin the leading party of France—but have at last acquired an appreciation of power and office, summed up in 1946's loudest, largest political demand, *"Thorez au Pouvoir."* Blum's scholarly humanism, his love for France, which has been more like a political romance than an expression of professional patriotism, his now weak and outdated Socialist Party (its once potent "Mystique of the Left" has moved east), his experience with the hatred that in his youth sent one French Jew to Devil's Island and in his old age sent millions of Jews of many nations to barbed-wire camps—all these elements, embodied in one leading Frenchman, seem finally to have completed their orbit. Blum's disappearance from the French political scene will be, on a small scale, like the disappearance of Roosevelt, except that there will be no funeral. A recent poll of the country showed that thirty-four per cent of the population is still devoted to General de Gaulle. He and Blum both now seem to be far removed from the governing of their land. But they are the two most popular men in France.

Politically, everything here moves sidewise rather than forward. Calculated horizontally, French Socialism, whatever it seemed like in the old days, is now practically the dead center, or pure middle-of-the-road democracy. The Socialist Vincent Auriol has just been elected first President of the Fourth Republic. While the election was going on in the Salle du Congrès at Versailles, many a rich man in Paris had his Versailles man on the phone for reassurance that the chances were favorable for Auriol's winning on the first ballot. And when he came in victorious, Paris big money probably broke open a good bottle of wine for dinner in relief. However much Socialism means to French capitalism as a breather, Auriol's election means to the already weak Socialist Party that a good active Party man has now become a eunuch politically. The new French President has even less political power than the presidents had before the war

The voting was dramatic. Just as some people always cry at weddings, some people in Europe now feel that it is a good time to cry at elections—with joy, to be in a place where mankind is still casting a free ballot. For the handful of Americans packed into the galleries, it was a sobering lesson to see the many women deputies calmly treated as if they were deputies and not women, and equally impressive to see the Negro deputies from Guadeloupe and other dark colonies striding with simplicity up the little staircase that led to the two green faïence electoral urns in front of the Speaker's desk.

The mounted horse guard, their trumpets playing, trotted the President's car from Versailles to his new residence in the Elysée Palace in the Rue du Faubourg-St.-Honoré. A fringe of French people lined the streets to watch the modest Presidential procession, but few cried *"Vive la République!"* The people seemed satisfied but quiet. France must wish to remain a republic, since this is her fourth use of the formula.

For six days, through this Tuesday, Paris was without French newspapers, because of a strike about which nobody knew the facts, because there were not any newspapers to give them. To fill in, the M.R.P. Party shrewdly put out a miniature, throw-away newspaper, its back page solemnly carrying a few square inches on horse racing, Admiral Byrd in the Antarctic, and food rationing, as well as a midget column called "News of the Entire World." Its front page was a scoop the size of your hand—the first excerpts seen in Paris of M.R.P. President Maurice Schumann's wishful-thinking speech on Germany, delivered at the Cleveland Council of World Affairs. France, as a little Latin country now ranking only a cut above Italy, knows that she will accept what the Ministers' Conference decides in

March, in Moscow, about the Rhineland, the Ruhr, and the Saar, just as Italy has accepted what they recently decided, in New York, about Trieste.

The only spot on earth France can still act big is in Indo-China, where, even if she were not losing the war, she has already lost face. Ever since England agreed, however reluctantly, after the first World War, to President Wilson's idealistic platform of free determination for small nations and Ireland promptly stepped aboard, one aftermath of this second World War has been inevitable. What with the British Empire losing ground in India and Egypt, and with fighting in Palestine; with the recent revolt (under British influence, the French say) in the French colony of Syria, and with the war against the Vietnam Republic (under Russian influence, the French say); with the Dutch Pacific colonies in unrest; with mutterings audible everywhere around the palm-tree fringes of the commercial globe; with notions of freedom having spread a long way in one generation—the problem of what attitude the white races should take toward their own and other white peoples' colonies is the most awkward, most important philosophical question in Europe today. The French Left favors liberty for everybody in Indo-China except the French capitalists' Banque d'Indochine. The fellow-traveller journal *Le Franc-Tireur* the other day declared, "One can fight against organized armies, conquer them, and make treaties. But one does not fight against a whole people when this people rises above itself and decides to risk the utmost suffering." France has been fiddling around in Indo-China, or "has interested herself in this territory," as the Rightist editorials elegantly put it, since the time of Louis XVI, when he slipped an alliance round the Emperor of Annam. From Louis XVI to the first President of the Fourth Republic is a long time for the French to have been in Indo-China without making friends.

March 20

For a period of thirty-one days, ending this past Monday, the capital of France functioned without any French newspapers to give it news of itself, let alone of the rest of the world. Though theatre box offices suffered a sympathetic slump while the newspapers were not being printed, Paris night clubs were crowded with the kind of people who crowd night clubs the world over, no matter what may be going on outside the doors. The newest nightclub hit here is a man called Yves Montand. He looks like a young Abraham Lincoln, sings workingmen's songs, of which the best is "C'est la Grande Cité" (the Great City being the Paris of the factories), uses his hands passionately, as if he were making a political speech, and was the French Frank Sinatra for young working Parisiennes, who are more restrained and respectful than bobby-soxers, when lately he sang at two music halls, the cheap Montparnasse Bobino and the Boulevards' A.B.C. He is now out of his modest admirers' reach, at the costly, champagne-drinking Ambassadeurs. Montand is a protégé of the famous *populo* singer La Môme Piaf. His name is really Yves Livi, and he is really Venetian by birth, with an Italian-Armenian mother and an Italian-Canadian father. Before he became a singer, he had been a barber, a dock laborer, and a factory hand. He is planning to go to New York soon.

The most acclaimed motion picture now showing on the Paris boulevards is the recently released Italian film "Sciuscià," pronounced "Shoosha," which, in turn, is the way Italian bootblacks pronounce "shoeshine," a word they picked up from the occupying American troops. "Sciuscià" is one more of those remarkable European films that have been made with a poverty of equipment and with rich emotion and are putting the slicker, emptier Hollywood productions to shame. The leading actors in "Sciuscià" are Roman bootblacks—a pair of incredibly talented, thin-legged ragamuffins picked out of real life on some shabby piazza. The film's love theme is their passion for a dapple-gray race horse, which they manage to buy through black-marketeering. They are arrested and sent to a Rome reform school, and the tragedy of the story and the precise, quick-paced artistry of the film lie in the fact that in this juvenile prison youthful events lead with the inexorability of the old classics to the poisoning of their friendship, to the ending of one boy's life, and to the ruining of the other boy's existence, no matter how long he might stay alive. The two boys undergo the complete human disaster that is usually harvested only by the adult man. This bootblack film provides the most informative, alarming, and concise picture of poverty-ridden, demoralized postwar European childhood that has yet been seen here.

* * *

For the past two months, there has been a climate of indubitable and growing malaise in Paris, and perhaps all over Europe, as if the French people, or all European people, expected something to happen or, worse, expected nothing to happen.

April 10

Until a fortnight ago, there was a popular after-lunch comic radio program called "Suivez le Guide" ("Follow the Guide"), done by the French music-hall comedian Max Régnier, assisted by Edward Stirling, formerly of the local Stirling English Repertory Players. Régnier acted as a French guide in current affairs, and Stirling followed him, in carefully slow French with a British accent. For saying nothing more than what Robert Schumann, Finance Minister, had already just said about the startling state of the French franc, though making it sound much funnier, Régnier and his program were taken off the air by the state-owned, state-censored Radio Diffusion Française. The program chief was thrown out, the Radio Diffusion chief angrily resigned, and the Paris press, as usual, burst into editorials, though since "Suivez le Guide" had been apolitical, Paris for once did not have to divide into opposite camps. Instead, it united in the old-fashioned and futile cry of "Down with the censors! Up with liberty!" André Philip, Minister of National Economy, said in a speech the other day in Bordeaux that "in reality we are menaced by a total catastrophe on both the economic and the financial plane." If Régnier gets back on the air, he should guide his comments away from these words. Nothing could make them funny.

Nobody can pretend that these are not sober days for France, as well as for other old seats of European civilization. Two months ago, it looked as if there were only one chance in a hundred that the Leftists would make bloody trouble in the streets of Paris, and only two chances in a hundred that the Rightists would make it; it also looked as if the chances were seventy-five in a hundred that for the next two years or so, while the United States, Great Britain, and Russia found out where they all stood, France would be able to worry along in her present strained but tenable condition. De Gaulle's Easter speech at Strasbourg has changed all the percentages. His success has worried everyone—except, that is, his followers. No one, not even he or they, can be sure what their newly formed League for the Betterment of France will bring—not so much with it as against it. On the whole, the French citizenry, from experience, is suspicious of a man who feels that he alone can manage history. For the past six months, France has been in the undignified position of an elderly lady doing the splits, her Right leg extended in one direction, her Left in the other, while everyone wondered how long she could hold it. The image called up by de Gaulle's league is prettier but more disquieting: the single savior on the white horse. Millions of French who had continued to admire de Gaulle, if only for his useful past, are now preparing to fear him—a painful, perplexing change of sentiment that will further addle and exhaust the country. The notable speeches he recently made at Bruneval and Strasbourg were spoken, respectively, on a picturesque hillside and in the shadow of a cathedral. According to the present laws of France, he will still have to talk in the National Assembly to make it count. The General's remarkably suggestive posters publicizing his Strasbourg speech showed France covered not by her hills and rivers and towns but by a photomontage of the sea of devoted de Gaulliste faces that had been lifted up to him when he addressed his thousands, al fresco, at Bruneval. Thus all France, according to his posters, is populated with his followers.

August 20

The Galerie Maeght has just held a show of Surrealist art, the first psychopathic aesthetic exhibition since the war. The de-luxe catalogue was idiotically priced at five thousand francs a copy. It was decorated with black velvet, on which reposed a plastic breast, the work of Marcel Duchamp. The show's japeries were stale *derniers cris*. In place of the prewar Dali taxicab with a shower in its roof and ivy as a passenger, there was, in one room of the gallery, a leaky ceiling that dripped upon a billiard table and the visiting art lovers. There were peepshows with tin cans and pebbles as the objects to be peeped at, and a fine Max Ernst painting of a phenomenal California rainbow. There were no Dalis, since he is apparently cross with the Surrealists, or they with him—though not so cross that they have refrained from copying his superior painting technique and paranoid perspectives as well as they know how. In the midst of the crisis in civilization, the intelligentsia Surrealists are still raking in cash through the inspiration of their private, patented muse—a composite of Marquis de Sade, Machiavelli, Maldoror, Narcissus, Mammon, and Moscow. Picasso has been more generous. He has presented ten pictures, painted in the past fifteen years, to the Louvre. The gift celebrates his happiness, in the autumn of his life, over the birth of a son last April.

The most interesting play of the theatrical season that ended last month proved to be "Les Bonnes," by Jean Genet, today's most discussed young poet and the newest member of that theatre-loving, inner artistic circle still dominated by the eclectic, lyric Cocteau, to whom the footlights have always served as evening stars. Genet's drama is about two tenebrous, visionary servant-girl characters who set out to murder their employer. Their dark plan is to poison the lady of their house, but it goes awry when the dominant domestic, through an insane form of social wish fulfillment, begins pretending that she, the slavey, has turned into the bourgeois mistress and that her sister is her servant. In her fit of *folie de grandeur* and double identity, she fatally serves poison to herself.

The season's most captivating operetta was equally dualistic at moments, but funny—the Opéra-Comique's "Les Mamelles de Tirésias," the ironic old Apollinaire fairy tale, now decorated with enchanting music by Poulenc. As the title implies, the bosoms of that classic heroine are major characters in the plot. Its climax occurs when they zoom off like balloons, whereupon Tirésias instantly grows a beard and, by parthenogenesis, becomes both father and mother of a stageful of baby soldiers, chorusing from cribs. Thus sterile, weak France, of which the heroine is a symbol, is repopulated, and indeed rearmed, to music, and her beard is allowed to drop off, her bosoms come back home, and the audience goes home, too, delighted with one of the few doses of national criticism that have caused mirth.

The same cannot be said for Albert Camus's new novel, "La Peste," whose important, dreadful theme is today's moral corruption and its spread through the social body. Perhaps it is a good sign of *mea culpa* that the Paris public bought up the first edition on the day it reached the bookshops. The reaction has been mixed. Those who enjoyed Camus's small, dry "L'Etranger" admiringly accept, in this broader work, the new protagonist pattern that both Camus and Sartre (increasingly the most influential pair of thinking authors since the war) have now set. In accordance with this pattern, human mediocrities are used, like a very common denominator, to decipher the problem of present-day life. It is their dateline rather than their dimensions that gives them importance. Mostly, they are people without personality, boldness, or the capacity to make shapely decisions—young citizens who are small, citified, ignoble, and monotonous, and who clamber clumsily among the little altitudes of their tragedies, which never seem higher than a Left Bank apartment house. The lack of tension in this modern school of fiction is old-style Russian, like Chekhov in the country; domestic minutiae are put down somewhat as they were in the nineteenth century "Adolphe" of Benjamin Constant, but the moderns' attention is on their organs, not their emotions. This new French novel style represents the greatest break yet with the Balzac style, though Balzac's most important theme was also the pestiferous corruption of French society, long before the Third and Fourth Republics put their hands into the grab bag.

1948

April 2

On the surface, France has improved so much in the last six months that all that is lacking is the belief that the vital, invisible underpinnings of state and society can hold the improvements in place. The average Frenchman can now find in the shops nearly everything he wants except the means of paying for it. In midtown Paris shop windows, perfect taste, which is the supreme French luxury, has at last reappeared. On the farms, the hedges have been trimmed and the ditches have been cleaned out, the farmers having profited from what may be the last year of German-prisoner farm labor. By day, in the limpid spring sunshine, Paris looks her old, beautiful self, reclining full length in the greenery by the Seine. By night, the city is lighted but largely deserted; restaurants are half empty and cafés are closed for lack of clients. Parisians dine at home, on soup, and go to bed. The rise in the birthrate here may be accidentally patriotic, but it is also alarming, considering the world shortage of food. Since the recent freeing of the franc and of the prices of most edibles and other goods, the frantic black-market bustle for necessities to eat and wear has quieted down. Paris is hovering around a new norm; the new heavy tax increases, the special fiscal levies, and the seizure of five-thousand-franc notes have combined to produce an odd, un-Gallic stoicism that is a substitute for morality. The anti-inflationary government, still insisting that the sellers' margin of profit is too wide, is using pressure to hold retail prices down, and has made a first public example of the vegetable marketman. In the spinach basket in his grocery shop or on his lettuce cart by the curb, he is forced to placard his cost price as well as his selling price. The fixers and the middlemen in the gray market—the nibblers rather than the producers—are the only business groups making big money. Factory workers say that wages must go up; factory owners say that prices cannot come down any farther without economic suicide. For political reasons, both sides exaggerate their difficulties, which are nevertheless real, and which are opposite. It is this absolute diametricism that is cutting France into two bleeding, anemic, and perhaps impermanent parts.

Many important items are still rationed. Except for doctors, taxi-drivers, and other specialists, the French do not get a drop of gasoline. Because the government acutely needs tourist dollars, gasoline flows in countering for American tourists, as well as for abashed American journalists (French journalists get nearly none), for whom the liberated franc makes life nice and cheap anyway. The French operate on costly black-market gasoline coupons that trickle down from the Brittany fishing ports, where fishermen make more money by selling them than by putting out to sea in their motorboats and catching fish. The farmers also oblige, by selling their tractor-gasoline coupons. Parisian adults have had no butter ration since Christmas and this month their quarter-pound monthly coffee ration is to be skipped, but they received a government Easter present—a rationed tin of sardines, at thirty times the prewar price. Wine is finally unrationed, but some of it is watered, or what the French call baptized, wine and spoils quickly. Compared with conditions a half year ago, there is more choice in everything and more comfort everywhere, except in the average French pocketbook. France is in a curious, momentarily excellent position of recovery. The goods are here, but the marts need faith and buyers with cash. There is a lot of intelligent skepticism as to what tomorrow and tomorrow will be in France, in Europe, and on this earth. France is like someone who has unexpectedly climbed a very high hill and stands breathless and poised on the crest.

French anxiety about the coming Italian elections is so grave that Italy is being spoken of with respect, possibly for the first time since the civilized Julius Caesar conquered all tribal Gaul. Only the Communists still refer to the Italians as *"Ces macaronis-là,"* a phrase they will, naturally, drop at once if the Italian Communists win. The French are also worrying about the possibility of their having their own spring elections, which could force the Communists and General de Gaulle's Rally Party into a final, or temporarily final, struggle at the polls, and perhaps in the streets. Lots of Frenchmen fear that if the General is elected, he will not be able to do much good, because retaliatory Communist strikes and sabotage could do so much harm. André Malraux, who, a fighter on the Republican side in the Spanish War, had been earlier admired for that great and Left-tinged book, "La Condition Humaine," on his youthful life with the Red revolutionary forces in China, is now the General's right-hand man. There is no left-hand de Gaulle man. At a recent private luncheon party here, Malraux talked brilliantly and rapidly and uninterruptedly for exactly two hours—or about twenty thousand words—on the General's political plans. He speaks with a strange, indirect tenacity, his sallow-skinned skull, with its sombre eyes, tilted to one side, apparently seeing nothing but his own thoughts, the expression of which is punctuated by his constant, nervous, maladive gestures and sounds—his dry little cough, his rubbing of his nose, his shuffling of his shapely feet. The main points he made in his remarkable monologue about his and the General's beliefs were these: The French bourgeoisie is finished; there is only one test for a class's fitness to rule—its ability to wage war; the French bourgeoisie did not defend itself in 1940; though the General draws his support from the Right, his political platform is for the workers, who, as all intelligent men now know, must have a better life; within six months, or perhaps only three months, after the General comes to power, he will have to have succeeded in giving the workers specific advantages, which must be so evident to everybody that even the Communists cannot deny them; and the Right will have to aid him in maintaining these working-class ameliorations. If de Gaulle could not do all this and in this period of time, Malraux said, he and the General would deserve to be shot, and probably would be. He figures that the unchangeable hard-core Stalinist French Communist voters amount to only six per cent of the electorate, and that there is an additional twenty per cent of what might be called fluctuating Communists, who are already worried about the Muscovite terror of liberty. The General believes that he already has forty per cent of France's voters and that some of the remaining thirty-four per cent, or the shilly-shally voters, can be drawn to him, and that, besides, things are weighted in his favor because he has the Army behind him. The east, including Alsace-Lorraine, is overwhelmingly for the General. In the west, he claims a majority. In any Communist insurrection, Marseille would become the Communist capital; the General must try to gain in the south. He is speaking at Marseille on April 18th, the Italian election day.

A new Sartre play, entitled "Les Mains Sales," is opening in a few days at the Théâtre Antoine. According to reports, it should restore the author to the pontifical level from which he has slipped lately. It is said to be a story of politics in the mythical Balkan country of Ilyria, where a young idealist assassinates his idolized political chief to prevent him from soiling his hands with ignoble political exigencies, only to discover that the government itself would have wiped out his chief, and thus kept the youth's hands, too, clean—in his case, of blood. Because of the times we live in, Sartre's "The Soiled Hands" will doubtless be no less popular than "Le Procès," André Gide's sonorous, impressive adaptation of Kafka's bitter myth "The Trial," now playing twice a week in Jean-Louis Barrault's Théâtre Marigny repertory. This is, visually, a rich, imaginative production, a synthesis of various somnambulistic states of mind. The finale is startling; Barrault, as the small, confused, innocent M. Joseph K. is led off to his beheading while treading air, being held aloft in eternal space by the polite, strong arms of his two giant, top-hatted executioners.

The Czechs never rated the Prague-born Kafka as one of their own, because he wrote in the hated German language. "Le Procès" is now arousing a belated nostalgic interest among members of the city's Czech colony. Since the new Putsch, the radio has had a lot to tell them about what is going on at home, if they will only listen. The main propaganda programs are at night. At nine, an anti-Communist news report, in Czech, comes from the B.B.C. in London, once more a seat of exiles, telling what is happening under the occupation and what the world thinks of it. At nine-fifteen, a Communist talk comes on, in French, from Prague. One of its recent anti-American items was the bland announcement that the American occupation forces in Germany were about to ship the Sudeten Germans there back to their Czech homes, a piece of propaganda calculated to arouse fury and fear in both the Right and the Left Czechs. So far, letters coming into France from Czechoslovakia have not been censored. A Czech in Paris has just received a letter from her sister in Prague relating her nine-year-old son's report on his first class under a Communist teacher. He reported that the teacher had said, "Children, you all know that in America people live in holes dug in the ground and are slaves for a few capitalists, who take all the profit. But in Russia everyone is very happy, and we in Prague are very happy, too, owing to the government of Klement Gottwald. Now, children, repeat loudly with me, 'We are very contented and approve the Gottwald government.'"

On the first day of the Prague Putsch, there were heartbreaking telephone calls from Paris to Prague, mostly young people calling parents still brave enough to talk out loud for a few precious minutes. One Prague mother, sobbing, shouted to her son that she would lock the door in his face if he tried to return, that Czechoslovakia was lost for a hundred years, that he must stay away and make his life elsewhere, and farewell, farewell forever, oh, my dear son. Since then, the son has learned that he will be condemned to death *in absentia* as a traitor if he does not return to undergo his military service. His father has had to sign as guarantor of his son's good faith in this matter, and can therefore be taken as his substitute into the Czech Army or before a Communist firing squad. The staple newspaper photograph of the long queues of Czechs lining up to view Masaryk's body lying in state was printed large on the front page of many of the morning papers. The Communist *Humanité* ran the picture, in pygmy form, on page 3, next to the fiction serial.

May 26

The Season is well under way—those May and June weeks that before the war were given over by Paris to a concentrated flowering of its civilized talents for fashionable marriages, masked balls, steeplechases, gourmet dinners, and theatre galas, and to the budding out of ladies in fine new clothes, which were in themselves part of the rich annual industrial harvest that the city lived upon. This year, the Parisians are less important to their Season and the vacation months to follow than the American tourists, a hundred thousand of whom are expected and will be profoundly welcome. The tourists are regarded as shiploads of precious material not specified in the Marshall Plan. Tourist dollars are desperately needed, to an extent that only the finance ministers of Western Europe can calculate.

The imported or otherwise special theatrical entertainment

customary during the Season has again turned up, if not in quite the usual sedate form. The old Marigny Theatre, picturesquely squatting in its midtown park of chestnut trees, at the moment in ruddy flower, is now the city's most catholic postwar entertainment center. Following Jean-Louis Barrault's success there with the repertory including works by Gide, Molière, and Shakespeare, with tickets priced at less than four hundred francs, came a week of frenzied *jazz-hot* concerts, with orchestra seats at eight hundred francs and with the house packed as it never had been for "Hamlet." The event, called *"La Grande Semaine du Jazz"* and organized by Le Hot Club de Paris and patron swells like Princesse Amédée de Broglie and the Duc de Brissac, offered American Negro masters like Coleman Hawkins, Howard McGhee, and Slam Stewart. It turned out to be a struggle between the diehard New Orleans style and bebop. The latter, judging, by what most of those present could hear of it over the catcalls of the French younger set, lost. Among the new French jazz talents on display were Claude Luter and his Lorientais and the recent Riviera Jazz Festival's prize trumpeter, Aimé Barelli, the finest French jazzman to come along since Django Reinhardt.

Major literary news concerns André Malraux, still considered the most talented and modern-minded novelist in France. A new book of his has been published, but it is one that few can buy. It is called "Le Musée Imaginaire," is an exposition of Malraux's new theory of obscure, realistic, reproductive art, and costs twenty-five hundred francs, or a tenth of an average, worried family's monthly income. It is a de-luxe collectors' volume, with a hundred and fifty-five illustrations, ranging from a gold Greek mask of Agamemnon to a giant pink face by Rouault. The kernel of Malraux's contemporary, revolutionary theory is that the reproduction of art, via photography and printing, produces what he calls modern man's "imaginary museum." "Today," he goes on to say, "the history of art is what is photographable." To him, photographs "at last make art a world heritage and permit its inventory." He deplores our habit of museums, institutions only two hundred years old, which have given us—as well as him, up to now—"our relation with art. The nineteenth century lived on museums. We still do." He says that photography, as a recorder of life, "in thirty years passed from Byzantine immobility [in daguerreotypes] to the frenetic baroques . . . of modern motion pictures." One of Malraux's few almost positive statements is his definition of modern art: "It is the search through form for an inner scheme that then takes—or not—the form of objects but in which objects are only the expression." The most indisputable line in the book is a quotation from that experienced old aesthete, Picasso: "Who are the younger artists? I am."

June 14

Another crisis has just arisen, over foreign policy and the results of the London Six-Power Conference, in which many French feel that France figures not as one-sixth but as about one-twelfth. The recent attempt by the American Congress to renege on the plans of Secretary Marshall, a move that dazed all western Europe, reminds the French of the reneging on the ideals of President Wilson after his Versailles Peace Treaty. The United Nations already reminds the French of the League of Nations; the possible withdrawal of the Allies from Berlin reminds them of their own withdrawal from the Ruhr in 1925; any Anglo-Saxon desire to lift Germany out of the beggar class reminds them of the Dawes Plan, the Hoover Moratorium, and American bank loans to Berlin; and, most frightening of all, any plan for unifying western Germany instead of isolating each of its provinces reminds them of Germany's organizing genius, which Parisians can see at work again, as it was during the Occupation, simply by shutting their eyes, and can hear, like an echo of the rhythm of Nazi feet along the Champs-Elysées. No sooner was the Six-Power London accord announced than the Communists plastered the boulevards with a Party poster, which, for once, was generally popular: *"L'Allemagne d'Abord? Non!"* ("Germany First? No!") If Bidault and his foreign policy fall in the near future, they will carry Schuman and his Third Force government with them. This would probably lead within the year to a government of the Fourth Force—the coming to power of the mystique and person of General de Gaulle.

Two of the best-known new literary figures, Jean-Paul Sartre and David Rousset, have founded a political party, heaven help us— Le Rassemblement Démocratique Révolutionnaire—and a bimonthly one-sheet newspaper, *La Gauche,* which, coming from professional writers, seems poorly written. Sartre's political ideas are less clear, if more optimistic, than his novels. His talent, his scholarly mind, his French essence, and his hypersensitivity to Europe's dilapidation give momentary importance to his political hopes, which concern that still unresolved trio of human problems, liberty, work, and lasting peace. "Hunger," he says poetically in *La Gauche,* "is already a demand for liberty." His rather less concrete goal is that "consumers and producers . . . may be conscious of their democratic revolutionary humanism." Rousset, whose remarkable book "L'Univers Concentrationnaire" recounts his survival in Nazi concentration camps, is more harshly definite on what Le Rassemblement is not. It is not Marxist, Trotskyist, or Socialist, but it welcomes all men from such, or any, faiths who are disorganized by deceptions and by political play and who have almost lost hope that "by democratic conduct humanity can demonstrate that it is not necessarily dedicated to self-destruction and barbarism." The Sartre-Rousset party declares that it expects to collect, in the next six months, a hundred thousand followers. If words were all, its followers should number millions, from all over this earth.

A mere handful of a hundred and thirty upsurging intelligent Deputies of many parties, and including the president of the Foreign Affairs Committee, have formally laid before the Parliament a resolution demanding "the immediate meeting of a European constituent assembly, having as its mission the founding of the permanent institution of a 'Federated Europe.'" Whatever else happens in Parliament, now or tomorrow, this is the one vital political proposal of the moment, of the year, of the century.

June 23

The most worried, wearied, unthanked, and necessary public servant in any government today is its Minister of Foreign Affairs. He is like a mother-in-law—in the bosom of the family, yet not of it. Essentially, he is related to a world outside, a go-between harried by what the family thinks is its due and by what the neighbors say it deserves, which is invariably a lot less. In the difficult last two months, Georges Bidault, still in his forties, has started to become an old man. The cynosure of all eyes in Parliament during its recent foreign-affairs crisis, and the butt of most of its discontents, he looked, as he sat engulfed in his great chair of office, like someone whom history had indeed dilapidated. His shining black hair had begun to grow white; the small, lively dynamo that supplies his physical energy appeared to be running down. His three years of peace as France's leading diplomat of the younger generation seemed to have cost him more than the long years of war, during which he functioned as President du Conseil National de la Résistance, the bold and successful chief of the underground forces in France.

The recent political crisis was the most protracted one in the history of the Fourth Republic; actually, it was composed of a series of smaller crises, each of which bubbled to the Parliamentary surface and broke in an explosion—if only of talk—that warned of national anxieties and agitations. As the talk grew more tense, the daytime hours did not suffice for it, and Deputies took to arguing all night and voting at dawn; in all Paris, for those hours, only Parliament and the Montmartre night clubs were up and hard at it, with some of the city's too few taxis working both stands, carrying the dissimilar, tired stragglers home. A typical Parliamentary rebel against the London Conference decision to put western Germany back on her feet (which was, of course, the main cause of the French foreign-affairs crisis) was the antique, Right Wing republican Louis Marin, an experienced anti-Teutonite who also rebelled, in his heyday, against France's accepting the Versailles Treaty. Of the London Conference plan, he cried, "We French are asked to sign it with our eyes closed, like the caged nightingales that are blinded so that they will sing better." Parliament was practically unanimous in its two classic Gallic fears: fear of France's naked lack of military security, and fear that control of the Ruhr, where in the past Germany has repeatedly rearmed right under everybody's nose, will probably be inadequate. The French feel that, on these matters, the Anglo-Saxons brushed France aside like the tertiary power she has become precisely because of German aggressions. There is no question but that the French have a fixation on the Germanies of 1870, 1914, and 1939. Only one Deputy gave precedence to the danger from the Soviet machinations in ruined Germany. "The London accord," he declared, "raises the fearful silhouette not only of a new German Reich but of a German-Russian Reich." The grimmest and most up-to-date Parliamentary comment was made by Paul Reynaud, surely an expert on German aggression, since he was Premier of France when it fell in 1940. Reynaud briskly begged his compatriots to try to forget Bismarck, and even Hitler, "because the atomic bomb has changed all that."

Ever since finishing its debate on the Conference, Parliament has been occupied with the inflation and the recent series of Communist-led strikes in protest against it. As the leading country of the remains of Western Europe, France is now trying to face up to a problem created not only by the second World War but by the first. France's inflation really began thirty years ago. Apparently, something is going to be done about it, at last—doubtless something that will hurt. In all the intervening years, no political party here has ever been courageous, patriotic, or sadistic enough to hurt the French voter in his pocketbook. Today, from the Skagerrak down to the Mediterranean, only the Swiss franc and the Portuguese escudo are hard money. It would seem that modern men have been smart enough at making money but that now they can't control the delightful stuff. French industry is producing at such a high rate that even Americans are investing in it. Business, however, is bad. Rich people can make more money, but lesser citizens cannot afford to buy or are hesitant about buying because they think prices may go down. France's fight against inflation will be waged with enormously increased taxes, with some magic schema that will hypnotize the French into paying them, and, most important, with austerity. This last, up to now, only the British have had the national character to support on the necessary great, dreary scale.

The vanguard of the biggest influx of American tourists since the great season of 1929 is now becoming visible and audible everywhere in Paris. They are as welcome as they are valuable. For everybody's pleasure, Notre-Dame, the Place de la Concorde, Ste.-Chapelle, and other architectural gems are once again being given their prewar evening floodlighting, which makes them stand out like well-proportioned ghosts of history. In the windows of the fine shops in the Faubourg St.-Honoré, the merchants, in a joint effort, have set up insouciant still-lifes illustrating the seven deadly sins. Two of those illustrating the deadly sin of greed feature fresh bananas and oranges. The most artistic is one in Lanvin's window, audaciously illustrating the sin of envy with a headless, and therefore brainless, female figure elegantly attired in court brocades and decked with real jewels. On the Left Bank, where entertainments are simpler, the St.-Germain-des-Prés quarter has become a campus for the American collegiate set. The Café de Flore serves as a drugstore for pretty upstate girls in unbecoming blue denim pants and their Middle Western dates, most of whom are growing hasty Beaux-Arts beards. Members of the tourist intelligentsia patronize the Rue de Bac's Pont-Royal Bar, which used to be full of French Existentialists and is now full only of themselves, often arguing about Existentialism.

July 8

The Grande Nuit de Paris, which topped off the Grande Semaine, was magnificently handled by government and municipal functionaries, against whose inefficiencies the citizens rail during the duller weeks of the year. Everybody coöperated except the weatherman, who produced sprinkles for the preliminary outdoor fête, which centered around the Tour Eiffel. The tower, floodlighted for the first time, looked like a solid-silver pillar of lace rising into the sky. Down on the earth and among the tower's four feet, trained elephants performed, along with other distinguished acts from the Cirque Bouglione. The star of the lot was an elderly matron named Maria, who, the Bougliones swear, was the favorite elephant of France in the reign of Louis XV, whom she has presumably never forgotten. At two o'clock in the morning, a burst of fireworks was launched from the Pont d'Iéna, below the great amphitheatrical setting of the Chaillot Palace. Earlier, in the Chaillot Theatre, at modest admission prices, there was vaudeville, with chorus girls from the Folies-Bergère acting as ushers. For people who could afford three thousand francs for a 2 A.M. supper in the heavenly upper reaches of the Tour Eiffel, the entertainment included Hollywood stars, among them Edward G. Robinson and Ingrid Bergman.

That night, all the historic monuments of Paris were flood-lighted, and Napoleon, atop the column in the Place Vendôme, shone like an enormous lamp on what is left of Ritz society. Autobuses, which nowadays normally stop running at nine, for the first time since the Occupation ran until two. Even the Métro stayed up, staggering around under the city. Cafés had special permission to keep open until two and to open again at four, for late revellers. For all the Grande Semaine activities, including its Grand Prix race, its evenings of ballet at the Opéra, its lighted fountains in the gardens of Versailles, pouring beauty over the night, and its elegant social affairs (for those who could still afford them), Parisians emerged from their worries and their homes; they spent some money, were cheered up, and, for once, had a good, prewar time. Today, more than ever, government is a form of applied psychology. Hope has to be advertised to the people. The Grande Nuit de Paris was a good advertisement of what Paris used to be like and may one day be again.

August 11

Even in Paris there has been excitement over the summer's humid hot weather, because of the prodigal wheat crop it ripened throughout the land and the hopes everywhere for that long-unseen luxury, white bread. Nevertheless, in view of the stormy political climate in and outside France, Paris, as the capital of what is still the Fourth Republic, seems tired. Lately, the Fourth has looked frail enough to fall, as the three earlier Republics fell. The timid, unpopular new Government of Premier Marie has led, owing to his pious name, to rude, contemptuous puns, such as "the government of the immaculate deception." The Schuman Government that preceded it was nothing but a Parliament-sitter, keeping an eye on the coalition of the little parties while the four big ones—Communist, M.R.P.-ist, Socialist, de Gaullist—carried on their doctrinal fights in real life outside, among their four kinds of French citizens, each of which, since liberation, wants France to be run in an ideological pattern that cannot be reconciled with the ways of the three others. This is not logically possible; this is why the Fourth Republic's survival has been in actual danger; this is why Finance Minister Paul Reynaud's eruption into the scene with his demand for exceptional powers—to save the franc, which, after all, everybody lives on—is also temporarily saving the Fourth Republic. In himself, though physically no bigger than a mouse, he possesses the power of the catalytic agent. As Finance Minister, he can go on being himself—before the war the smartest tax collector of modern France—and the four ideological parties can go on being different, attacking each other and also him, the tarnished Premier of 1940, who handed over the fallen Third Republic to Marshal Pétain. Today, Reynaud sits with idealist, Socialist Léon Blum strangely at his side. "We must save the Republic," the wise old Talmudic patriot said—with the truthfulness that has lately helped ruin his Party—adding as a whispered Party afterthought, "for without it Socialism is impotent." His Socialism has now lost its prestige, just as it has lost its working-class cohorts and voters, who slipped off elsewhere, to Left and to Right. Over the last year, as if today's Parliament were Utopia, the Socialists uniquely, under Blum's idealism, have helped everybody but themselves. The French Socialist Party's present decline is a loss in moderation to all Europe.

October 5

General de Gaulle is very pituitary these days, to judge by his increased appearance at his recent, and important, press conference. It was appropriately held beneath crystal chandeliers, in the elegant old Rue François Premier mansion that has become the headquarters for his French Rally Party. Time, weight, and, evidently, the General's glands are giving his visage a heavy, royal outline; he looks more like a man of dynasty than of destiny. His military voice, which sometimes soars to a treble, and his finical, old-fashioned phraseology make him seem like an eccentric middle-aged monarch. In certain respects, he seems to have improved. There is a closer, quicker, more dynamic connection between his ideas, his words, his remarkable memory, and his convictions. The sense of disaster in present French politics, and the General's own special apocalyptic nature and his growing following, now estimated at forty per cent of voting France, made the content of his press meeting, transmitted by him to us journalists in his special party jargon, headline news. He and his party refer to his followers as "the Companions" and to the Communists as "the Separatists." "The Legality" is their name for the government he would head when and if called to power "by the people's will." He told the press that the Separatists could have representation in his Legality but could not actually have any vote, lest they turn it into what he calls "the Illegitimacy." He used the word "Association" to describe his plan for coöperation between capital and labor, adding that workers must be given an interest in production. Nobody knows whether this means profit-sharing, a fact that may indicate that his main backers are not—as his Leftist opponents claim—the rich industrialists. They would, presumably, long since have rassled some sort of definition out of him on this vital point. For the rest, he was vehemently, intelligently explicit about what has been wrong with every government since he ceased governing.

The General claimed that his party had already realized ten million francs on the sale of its fifty-franc stickers, or stamps. The stickers, an idea of his propaganda chief, the novelist André Malraux, show two Lorraine crosses below the female bust and arms of Rodin's winged Republic, part of a larger group designed as a monument for Verdun. Other political groups have had a lot of free fun with the General's stamps. The Leftist newspaper *Franc-Tireur* prints each day a parody stamp, depicting the General with an umbrella for a body, accompanied by the request that readers cut it out and send it to the General's headquarters, which thousands have done. The Communists' excellent weekly intellectual magazine, *Action*, printed a stamp, in three sections, showing the tall General snipped into three parts and lying flat on his back. Jokes aside, the Parliament's decision to have no major popular elections in the autumn and merely minor ones next spring means that the only way de Gaulle can soon come to power would be by a *coup d'état,* with his hypothetical forty per cent of the French voters at his heels. Against them and their determination to save France by what they are convinced would be an honorable republic would rise that third of France made up of Separatists, who are determined that France become what Moscow now calls a Socialist democracy; i.e., Communist. The French Communist Party's political bureau has just announced its latest Moscow order and slogan: "The French people will never make war against the Soviet Union." In case of a de Gaulle *coup d'état,* the French people would probably fight one another. France is the only European country that is talking about a possible civil war. It's a change, anyway.

France is in a dangerous condition. Her strength is flowing away in choleric politics and falling money. Paper money is worth no more than the total of the national morality behind it. Only voluntary unselfishness and farsighted sacrifices can really help. In the end, France will probably go bankrupt, which will, at any rate, settle things. The French franc is now back on the dollar black market and underselling itself by a clear third. There have been strikes in French mines, banks, government offices, and transportation, gas, and electricity services. At the moment of writing, sections of Paris are without water while the waterworks workers are spending the day protesting against something. Within the month, high prices have become fantastic and, as always, contagious. In the countryside, fertilizer has jumped twenty per cent. In Paris, a ride in the Métro has jumped a hundred per cent. Squeezed by these increases, French life stumbles along.

November 4

The dramatic, unexpected turn in our Presidential election caused extreme surprise, gave great satisfaction, and certainly aroused sentiment here. According to the French calendar, the election took place on Le Jour des Morts (All Souls' Day), a day on which the French celebrate the memory of their dead. To Paris, the election of the Democratic Party and President Truman was a gesture by millions of Americans in remembrance of Roosevelt. For the average citizen here, the main question about the two candidates and their platforms—ideologically indistinguishable to most Continentals, now expert only at evaluating political extremes—was which of the two, the familiar Democrat or the new Republican, would do what for—or with—Europe. This is the attitude of Paris, or any other Western foreign capital, toward Washington today—an acute, vital curiosity as to what our dominant political personalities and elected bodies may decide about the rest of the world.

December 24

France has not fallen during 1948, perhaps less to the surprise of the French than that of everybody else. Looking back on the year, this is the main good news. France apparently cannot knock herself out; she can be knocked out only by some outside force. Though battered by the body blows of her own politics and some of her own people, France is too well fleshed—her land is rich—to do more than stagger sometimes and worry everybody. But she does not fall. The chaotic politics of her republics have not killed her yet, and she is now on Republic No. 4. Unfortunately, what this fourth one has achieved is the logical paradox of twentieth-century European democracy, a mixture of freely elected parties—roughly Left, Center, and Right—in sufficient balance to paralyze one another. The chief political fact of 1948 is that this set trilogy has wavered somewhat. The Socialists, wandering too close to Center, have nearly collapsed, and that is the year's political tragedy, for they are the natural New Dealish bulwark against both reaction and Communism. Communism's supposed loss in popularity is equally important, if equally true. One way or another, France's middle, and muddling, Third Force government has grown a bit stronger. At General de Gaulle's recent monster mass meeting in Vél' d'Hiv, he and his Party chiefs sat high up in a central tribune, with his lesser chiefs sitting below and his rank and file sitting lowest of all, in an obvious hierarchy. If he should come to power in the 1949 spring elections, he could give the French a very odd, unrepublican republic indeed.

1949

April 15

The Marshall Plan's European Recovery Program being one year old, Minister of Finance and Economic Affairs Maurice Petsche recently made a happy-birthday speech over the radio. It was addressed to the United States, but it was really intended for the ears of the French nation. He was explicit; he thanked the Americans for the millions of tons of coal that have kept the wheels of French industry turning; the cotton that the French mills weave two days out of three; the wheat that has lately supplied a quarter of the French bread; and the gasoline on which French trucks roll one day out of two. "All this merchandise," he said with emotion, "has been given us gratis by the American government. A great lifting of the heart goes from us toward the generous American people and toward its leaders."

In this fourth year of peace, France, along with its neighbors in Western Europe, has definitely turned the corner and is rattling down the highway that always had, still has, and, democrats hope, will continue to have some deep political ruts, since only the Communists here guarantee to steam-roller everything smooth. It is now expected that most of what remains of the black market, which has persisted ever since the Occupation, will disappear before the end of this month, when milk, butter, cheese, and maybe fats, oil, and chocolate will be available without ration tickets—at a stiff boost in price, naturally. The franc is soaring, the dollar slumping; there is talk that unloaded black-market dollars may even fall below the legal rate. The Quai d'Orsay's Economic Department of Foreign Affairs has asked the neighboring countries to be ready on May 1st with their money-stabilization plans. American journalists here say they are filing only sixty per cent as much news as they were sending from France a few months ago. This is itself banner news. What it implies is that France is now forty per cent less a headline worry to herself and others than she was when this fourth year of peace began.

Jean-Paul Sartre's Existentialism has often been attacked as a philosophical underminer of postwar French morale, but his latest work of fiction, "Death in the Soul," now being serialized in his monthly magazine, *Les Temps Modernes,* is a novel of high, anguished patriotism. In it, he grieves, like an author and a man, for the loss of one he loved, with all her faults; he grieves for France after her rapid fall in 1940. The novel, the fourth lengthy installment of which is appearing in the current issue, is the third book in "The Road to Freedom," a projected tetralogy on France and some of the French in the past decade. "La Mort dans l'Ame" opens with Sartre's familiar, confusing horizontal technique of aligning unconnected characters and early and late patches of the same event. He starts with a Spanish Republican, in New York, glad that France has fallen; a young White Russian, in Cannes, and his decision to join de Gaulle if his cancerous elderly French mistress will let him; and a Parisian defeatist's bitter pleasure as he notes the handsomeness of the German army marching into Paris. It is only when Sartre backtracks in time and finally focusses on some aimless French soldiers, recumbent and alfresco somewhere at dawn, waiting for day and defeat, that his perspective contracts, and then he stands over his characters with a magnifying closeness. "Where are we?" he asks, speaking for them. "In the grass. Eight citizens in a field, eight civilians in uniform, rolled two by two in one Army blanket, in a vegetable garden. We have lost the war. It was confided to our care and we lost it. It slipped through our hands and was lost somewhere in the North, with a sound of breaking." The schoolmaster Mathieu, from "The Reprieve," the preceding novel in the series, is among this eight, for whom the entire Battle of France consists of exactly fifteen minutes of useless fighting.

At least in its first four installments, "Death in the Soul" is often as dull and as talkative as life, with the virile grossness of speech that is typical of the last war's war books in both the United States and France and that makes many women readers in either language feel civilian and pacifist indeed. But Sartre's new novel also contains enriching formulations and a civilized thoughtfulness that mark it as French. More than any other French writer, he is credited with the sensitivity, talent, philosophy, and patience required to reduce modern history to the size of the individual man.

The artist Christian Bérard recently died—as he had lived for many years—while working, after midnight, in the Marigny Theatre, which houses Barrault's Repertory. His décor for Molière's "Les Fourberies de Scapin," now on view there, is, in a way, his Transfiguration Scene. The single set is a perfect example of his precise sense of stage architecture as a three-sided public shelter for well-rehearsed emotions and actions, with his solid constructions given imaginative perspective by his characteristic, artfully blurred, painter's details, which have always made his scenery look like something seen far off and through time. He was an Ile-de-France painter, and a climatic Paris gray was his favorite color. The décor of "Les Fourberies" is gray, as are the costumes, which are so elegant that they look like portraits of period clothes. Bérard was an authority on Molière and was France's greatest living theatre artist; he more and more used the stage as his easel, to the detriment of his studio canvases. Of late, his attachment to his infrequent paintings, even when he was working on commission, was such that he was inclined to think that they were never finished, so he kept them with him against the day he would add another loving brush stroke. He was a mature eccentric even when he was a good-looking, unbearded youth, and at his death, when he was forty-six, he had become a Paris legend of untidiness and fashionable taste. He was a painter who made his own style and influenced his own time.

April 28

Josephine Baker has returned to the Folies-Bergère as its leading lady. She is still excellent French star material. Her voice continues to be as sweet and reedy as a woodwind instrument. At the beginning of her long Paris career, she looked Harlem; then she graduated to Creole; she has now been transmuted into Tonkinese, or something Eastern, with pagoda headdresses, beneath which her oval face looks like temple sculpture. Her show consists principally of her changing her costumes, which are magnificent. The finale is exceptional. At the top of a high purple staircase, as a well-dressed Western Mary Queen of Scots, wearing a white train that stretches from the flies down to the footlights, she is beheaded, and from behind a veil (to hide her loss) she immediately sings Schubert's "Ave Maria" (written more than two hundred years later), in Latin. Then the whole theatre is turned into a French Gothic church by the appearance of phosphorescent stained-glass windows, which glow above the boxes and the balcony, while the stage is filled with phosphorescent stained-glass-window figures—a regular Folies-Bergère chorus from the clerestory of Chartres Cathedral. This Scottish finale is probably the most spectacular anachronism ever seen on a Paris stage.

May 11

Visitors who want jazz-night-club life will certainly find it on the Left Bank, in the St.-Germain-des-Prés quarter, which has become what Montparnasse used to be. Claustrophilia is the St.-Germain fashion, with the *boîtes* in basements—some of them authentic eighteenth-century cellars that are still unventilated, after all these years. The best new club of the kind is in the basement under the Vieux-Colombier Theatre, with the gifted Claude Luter's *jazz-hot,* and black French African natives singing home songs. The Rose Rouge Club, on the Rue de Rennes, features Les Frères Jacques, a troupe of talented young actors who have become the star entertainers of Paris. Their singing and pantomiming, in sweaters and top hats, of the grammarian Queneau's recent French literary classic "Exercises in Style"—ten ways, tough or refined, of relating the same silly story about a man with an unusual hat on an autobus—is very comic. The Club St.-Germain and the Tabou, the first of the troglodyte caverns called Existentialist—meaning merely jive—are still feverishly popular. There are also Sunday-vespers jam sessions at the old Right Bank Bœuf sur le Toit, now on the Rue du Colisée.

It is sad to have to announce the death of Louis Moysès, known to *tout Paris* as Moïse, founder and patron spirit of the various incarnations, on various streets, of the Bœuf, of which that on the Rue Boissy-d'Anglas during the fabulous nineteen-twenties and early thirties was the celebrated peak. This Bœuf collected the most brilliant figures of a brilliant epoch—the best-known international painters, writers, poets, poseurs, actors, and eccentrics; English milords, dressmakers, and dilettantes; the most talked-of women; the most openhanded millionaires; and the most entertaining parasites. Dadaism, Cubism, and Surrealism all flourished there, amid evening clothes. One member of the Bœuf's famous piano-playing team, Wiener and Doucet, used to read Rimbaud from a book propped on the music rack while he whacked out early boogie-woogie variations on Bach fugues. All that was long ago. One of the Bœuf's writers was tortured to death by the Gestapo, one perished as a slave laborer in the bombing of Hamburg, one committed suicide to avoid being shot as a collaborator of the Nazi Otto Abetz. The Bœuf poets have become bestsellers or Communists. Its greatest painter has become an international political figure. Many other habitués have merely become settled and stout. After the Boissy-d'Anglas's establishment, the Bœuf ran downhill. Few Paris newspapers mentioned the death of M. Moysès, so only a handful of the old Bœuf hedonists rose early enough to go to his funeral, at 9 A.M., in the Neuilly cemetery.

May 26

After a month's refreshing vacation, the French Parliament has just reassembled to deal with a couple of tedious problems that have plagued France intermittently for centuries: how to get the money to balance the national budget and how to stop a war. The war against the insurgent natives in French Indo-China has been dragging on for two and a half years, practically ignored by French politicians and little noticed by most other French citizens, unless, of course, the citizens have had sons fighting, and perhaps dying, there.

As for the war in French Indo-China, *Le Figaro,* politically ever the arch-conservative, has just admiringly stated that Annam's ex-Emperor Bao Dai, who is obligingly pro-French and who recently returned to the country as a mediator from voluntary exile, "conducts his political action like a sportsman." The Socialist *Populaire* and many middle-of-the-road French frankly favor the fighting national-ist Ho Chi Minh, who wants the French Army to get out and stay out, and the colonials to be independent. The Communist *Humanité* is loyally for the Communist rebel government, is espe-cially against the Banque d'Indochine (and Wall Street), which it says are at the bottom of everything, and damns what it calls the Dirty War—*"La Salle Guerre."* Before the Parliamentary vote on the weighty Indo-Chinese matter—when it was decided that Cochin China and Annam and Tonkin be joined to form the new state of Vietnam, a decision on which a French Indo-China truce may yet be wangled and the war stopped—only fifty Deputies bothered to attend the debates. One of the fifty (who is also a noted professor of anthropology) wearily and wisely remarked, "In any case, in a few months the Communist Chinese armies will have reached the northern frontier of Indo-China."

Prices are dropping in shops and restaurants. So is the recently boosted franc. In 1926, the recalling to the premiership of the strict Raymond Poincaré started, within a few hours, a stiffening of the franc toward a value (twenty-five to the dollar) that it maintained, with only occasional fluctuations, till the beginning of the next war. Today, though, probably only the minting once more of the pre-1914 continental silver coins could check the European inflation that two wars have produced. With astounding ingenuity, France has pieced together a new prosperity, based on a falling paper money. Big business is coming back into politics, and politics is again becoming a cloak for shrewd horse-trading combines, operating under new names. The small, smart, city-type Paul Reynaud, now head of the Independent Republican Party, and Pierre-Etienne Flandin, who was Minister of Something-or-Other in everybody's Cabinet, including Pétain's—these, along with similar long-dormant figures in the discredited Radical Socialist Party (that well-to-do manufacturer of politics and policies for France during most of the Third Republic), are jockeying to get back into the seats which they held so competently, within their limitations, and out of which the convulsion of the war finally threw them. The new French political look of 1944-45 has, understandably, been lost in the revival of bourgeois prosperity, which was thought necessary to fight France's share of Communism in Europe, and the French reform parties that came out of the war searching, as usual, for a brave new world must now, to keep alive, look around for ways of being practical. If the Third Force government falls in the present crisis, the elements out of which the Third Force was itself constructed, plus the more practiced, intense elements that have rallied around it, may well be able to repeat France's interwar phenomenon of the too famous *cabinets des cadavres*—government by the same deadpan faces, in changing official chairs.

It is a remarkable tribute to France that she must now be credited with having in four years, amazingly, recovered from the war. It is a pity that she seems also to have recovered from the spirit of the Liberation.

October 7

The summer is over, and the drought with it. The greatest harm the drought has done is to shake the French peasants' confidence in the necessity for rain. This year's sugar beets are, it is true, unusually small, but they contain an unusually large amount of sugar. This summer's Burgundy grapes are scarce, but they are especially sweet and will make another vintage wine—the fifth in a row. Fruits, notably pears, have been abundant; the lack of rain was made up for by sunlight and the absence of destructive winds. The wheat was more plentiful than it was last year, even though it was short-stalked. Potatoes did suffer. Ile-de-France farmers are digging only twice the weight of potatoes they planted, though an eightfold return is normal. And the farmers cannot compensate for the arid pastures of August and the one crop of meadow hay instead of the customary three. Cattle were slaughtered because they could not be fed. The French government, having turned groceryman, like all governments, has subsidized butter—which it calls "a demagogic victual"—so as to permit a three-per-cent price cut, to stifle the age-old political complaints, from the Left, of high prices. French cigarettes de luxe have just been cut fifty francs, or fourteen cents, a pack, and American cigarettes, at fifty cents a pack, in francs, will be sold in government tobacco shops beginning in November—too late for the tourist trade. The Queuille Cabinet's crisis was provoked by the Socialists' demanding both lower food prices and higher wages for underpaid workers, which workers generally are also insisting upon before prices gather strength for the new jump expected as a result of international devaluation. The cost of food has already risen. According to the official index, it was up four per cent in September.

Another Leftist political undercurrent was set going by the Communist Party's private "vote" for peace on its International Day of Struggle for Peace (Sunday, October 2nd), which culminated in a monster rally around the voting urns set up for the sabbatical balloting in the Parc des Expositions, near the Porte de Versailles. *L'Humanité* next day claimed (as usual) that the rally was attended by "hundreds of thousands of Parisians." The vote was a very successful propaganda sprout of spring's World Congress of the Partisans of Peace. Six Communist mayors were temporarily suspended by the Minister of the Interior for helping local voters to vote, and the Minister of Education publicly warned schoolteachers to eschew any connection with "this illegal referendum." The ballots were printed in September by the Party's elaborate presses and carried the words "I Vote for Peace." The conservative *Figaro* commented that to expect anybody to vote for anything but peace was as silly as to expect anybody to vote for a plague.

The British were the first tourists to make the rounds of the Continent, several centuries ago; then the Germans began travelling in hordes, accompanied by their Baedekers; and finally came the Americans, preferably to Paris, accompanied by their dollars. Without the Germans, with few British as compared to those of the *milord* days, and with only a hundred and sixty thousand invaluable dollar citizens from the States, the 1949 season has nevertheless been a spectacular one for European travel, with an estimated six million people—mostly Europeans—footloose at one time or another. There was also a rich mixture of exotic voyagers, such as Arab chiefs in linen robes, Hindu wives wearing diamond nose buttons, Chinese, Israelis, Turks, and an extra-heavy contingent of elegant South Americans. Even the upper-class French, who rarely go anywhere (except into occasional exile), have been travelling outside their own land. In preparation for an expected three million Catholic pilgrim-tourists to Italy next year (which is an Anno Santo), who will put that country at the top in tourism, Rome's mayor recently concluded an investigation into why his city has been running second to Paris. He discovered that the answer is noise. Noise was suppressed for a time, by Mussolini, and the Demo-Christian Republic has for political reasons feared to bring the matter up again, but it will now take steps. For tourists are the most cherished big business in Europe today.

Not only was the Paris tourist season the best in all Europe but it is continuing, thus turning into the longest one on record. Paris hotels, chic or shabby, Right Bank or Left, are still packed with summer leftovers. The Paris hotels, which, in comparison with those of London or New York, were always cheap, have gone up formidably in price, owing to taxes, wage scales, and the postwar French system of high profits to replace lost capital. Last year, the Government Tourist Bureau feared that Paris's rising hotel rates would scare off tourists or at least make them scream. The only outcry right now is from frustrated travellers who can't squeeze in anywhere, regardless of what they are willing to pay. This summer's mass travel in Western Europe was probably a logical enough result of its recent history. For six years, almost nobody travelled except soldiers and those segments of the population that made an exodus in fright or in fatal, forced emigrations. Some people travelled then because they were ordered to, while others, shut in, yearned in vain to move about. And there were not enough trains, food, or, most important, money, all of which now seem to abound. It is difficult to believe that Europe could change so miraculously and become the great, pleasurable, money-making and money-spending touring ground that it has been this season, exactly one decade after the war season of 1939.

The liveliest revue in Paris features that oldest of old tourist favorites, Mistinguett, in the smallest, funniest, lowest-priced show of her endless career. She must be almost the age of the Third and Fourth Republics combined, and her famous legs have long since assumed the impersonal rank of public statuary. Her show, entitled "Paris S'Amuse," is housed in the little A.B.C. Music Hall, with tickets at from a hundred and twenty to six hundred francs—thirty-four cents to a dollar seventy—and with costumes that are also strictly cotton-grade. Mistinguett is an amazing sight; she still has her lovely, white, sharp-looking teeth, her handsome blue eyes are still empty of thought, and her voice still tintinnabulates on pitch, like a nice, weathered sheep bell. Her dancing has become a series of gingerly skips. To judge by the topical songs she sings about herself, she is still hard-working, still rich, and still growing richer. Big Paris revues such as those she used to star in were traditionally nude and dull. With elderly showmanship, she has seen to it that the sketches in her current revue have satire and good acting, and she has, further, arranged that her chorus girls be buttoned up to the neck.

October 23

Paris has too many automobiles for comfort and is about to have many too many more. During the last fortnight, the Municipal Council's Police Committee announced that there are now twice as many cars in town as there were in 1945, that they are producing the worst traffic bottlenecks in this ancient city's vehicular history, and that two-thirds of the injuries in the Paris Police Department are suffered by the unfortunate traffic squad; and the thirty-sixth annual Salon de l'Automobile has just opened at the Grand Palais to sell more automobiles.

Sixty-four canvases by Picasso, painted during the last two years, have recently been on view at the Maison de la Pensée Française. Since a dozen of these pictures, in his latest distorted manner, are almost recognizable as portraits of children, a list of his offspring is tacked on the gallery wall for the art-lover's information. According to this, his eldest son, who was the model for the lovely harlequin-boy portraits in the early nineteen-twenties, is now thirty and is not represented in this show. Nor is an eleven-year-old daughter. However, his new little son, Claude, and his baby daughter, Paloma, both by his handsome mistress, Françoise Gilot, repeatedly figure in it, as does the mother. A portrait of Paloma, "Infant in Her Perambulator," with the pram wheels and the babe's eyes and nose going off in unexpected directions, might well frighten any child beyond the pram stage into fits. There are two equestrian studies of Claude, both called "Child on a Rocking Horse." Three portraits of the children's mother show a new, more benevolent family style—a normal full-face likeness of a pretty and blond woman, overlaid by a kind of green curtain of traditional Cubist lines. There are also a few magnificently astringent still-lifes, one of which is the hit of the show. Its subject matter sounds absurd but looks both rational and poetic—a stuffed owl perched by a study table and clutching a long spear. In these still-lifes, the genius of Picasso's composition and line serves once more as the signature of the modern master.

At the city's Musée d'Art Moderne there is a retrospective exposition of ninety paintings and drawings by Fernand Léger, done since 1905. It covers all his periods, from cubes through mechanics, tubes, architecturals, dynamics, objects in space, and the American epoch. His latest is husky ladies wearing tights and riding bicycles. The major opus in this new vein is called "Leisure."

France has been like her old self again; that is, without a government for more than two weeks and torn by an angry, noisy, doctrinary, disputatious crisis in Parliament. The Socialists provoked it by insisting on higher wages for a small section of underprivileged workers. The Socialists then tried to calm things by proffering as Premier their own unpopular Minister of the Interior, Jules Moch, who was bound to be offensive to the Left, because last autumn he loosed the Army on France's striking miners—an act the proletariat still recalls as unforgivable. Probably the Socialist Party has stretched itself too far, it being now like a rainbow, with red radical members on one end and true-blue nationalists on the other. The third party in size for more than a year, it has been first in its parliamentary job, which is to serve as a willing minority that, when added to a minority even smaller, like former Premier Henri Queuille's Radical-Socialist Party, tots up (with a little extra help) to a working majority by which France can be governed. Georges Bidault's Popular Republican Party (M.R.P.), second in size, has been unable to get minority help, because it is considered too pro-Church. The Communists, who are the leading party, have also been unable to get minority aid, for reasons too many to enumerate. France has therefore been governed by minorities, elected by a minority of her citizens. Those who lost the elections have been running the country. When this anomaly blows up, as it just has, it takes time for it to be put together again.

November 4

During the better part of October, France had no government. Nor is it certain that things will be improved politically by the earnest new Premiership of Georges Bidault, of the Popular Republican (M.R.P.) Party. In his opening declaration to Parliament, he spoke sadly of "the indifferent eye turned by a tired nation" on democratic (and quarrelsome) institutions like the one he was speaking to. For, after all the fuss, he came to office with practically the same political program as that offered by the Socialist Jules Moch, the most bitterly rebuffed of October's "designated" but not "invested" Premiers. For a minute, it looked as if Bidault, an ex-Premier himself—from the provisional government set up after the Liberation—might have to make up his cabinet almost exclusively of other ex-Premiers, of many political varieties.

Parliament's expert cynics now prophesy that the moneyed bourgeoisie, strengthened by its recent astonishing recovery, is determined to crack its way back into political power. If it succeeded, the actual boss party would again be the Radical Socialists—in which businessmen are as thick as plums—the party that was longest in power in the prosperous heyday between wars and that brought, along with its valuable energy, get-rich-quick scandals and political atrophy. Since the war, the Radical Socialist Party has lain doggo, and time has washed away some of its stains. Surely the answers to the questions now before Parliament, which include the freeing of wages frozen in wartime, the renewal of collective bargaining, and a review of the privileges of strikers whose actions affect public economic safety, concern employers no less than employees. The French think that what France is going to have to face up to is another general election, and soon. Any regime between now and then may be merely a precarious government of day-to-day events.

A series of unflattering analytical anti-American articles, intended principally as merely pro-French, has caused surprise and chagrin in both nations' official circles here since it began to appear in the dignified *Monde.* As a revised, postwar version of the august *Temps,* this newspaper usually achieves, as a diurnal miracle, the accuracy of a lexicon in its writing style, a total absence of humor, and editorials as heavy as bars of gold. Pierre Emmanuel, the author of the articles, the subject of which was what he terms *"L'Amérique Impériale,"* is a youngish French poet and philosophy professor, whose first unsatisfactory visit to the United States apparently occurred, in the company of his parents, when he was aged four. As a polemicist, he has a manner that is an uneven mixture of insults and fiery critical judgments. In one *Monde* gibe, he coarsely remarked, apropros of our President's reported approval of the Vatican's excommunicatory decree against militant Moscow, "And nobody even cracks a smile on seeing the former suspenders merchant transformed into a defender of the faith. 'Hello, Pope!' says Mr. Truman." Emmanuel added, "For five years, American propaganda has been based only on fear [of Communism], a panic sign of impotence. Production-consumption is the diptych of American existence. America kills itself with work. It is a continent hostile to anyone who thinks. Publicity hammers the American brain just as propaganda does [in Russia]. Old countries like France, Italy, and England remind America of its inferiority complex." Emmanuel excoriated Americans for taking to drink and sex. Indeed, he said, America's interest in sex has been its most obvious donation to postwar Europe, where "even in the land of Stendhal, *le phallus est en passe de devenir un dieu*," a statement that must have astounded the polite *Faubourg* no less than it did the Marshall Plan administrators here.

Mme. Colette has recently been chosen president of the Goncourt Academy—the first woman ever to receive this literary honor. The ovation given her on the opening night of "Chéri," her own dramatization of her famous novel, at the Théâtre de la Madeleine was also unique. Elderly, arthritic, ensconced in a stage box, from which only her head was visible—her still mordantly witty face surrounded by its nimbus of radiant hair—she received the acclaim of what is left of the three generations of *tout Paris* she has known: of French government representatives, members of her own and other academies, the leading poets, writers, and artists, and less prosperous admirers, high in the gallery, who have loved her sad love stories of the froufrou days of gay Paris before the First World War. As for the critics, they paid their homage in their reviews of the play, even though the novel's sombre nuances and emotional flow were lost under the spotlights and the theatre's rigid necessity for three acts. As the leading, fading cocotte of the play, Valentine Tessier seemed more a fine actress than a convincing demimondaine. As Chéri, the old-fashioned gigolo, the romping Jean Marais appeared not to know the more romantic tricks of the trade.

Paris critics were, on the other hand, unanimously severe with the modern amoralities of Tennessee Williams' "Un Tramway Nommé Désir," produced, with the addition of odd blackouts and Negro dancers, in a very free French text by Jean Cocteau, at the Théâtre Edouard VII. Arletty, theretofore known as a comedienne, played the sad, mad Blanche. The piece is a hit with the public, at least. The farther this New Orleans play goes afield, via foreign-language productions in Rome, Brussels, Sweden, and

Paris—and even in the English of London—the less, apparently, its meaning resembles the meaning it had in New York.

November 16

In the small back room of a new Left Bank bookshop called La Hune, a James Joyce exhibition is going on. It is almost as weighty, concise, and personal as his greatest book, "Ulysses," around which it is, properly, built. Joyce died in a final exile, in Zurich, in 1941. His library, his manuscripts, and other possessions that were found in his Paris flat after his death, and were to be sold for back rent, were lovingly and illegally salvaged by his friends Paul Léon and Alexander Ponizowski, both of whom later died in a concentration camp, after being deported by the Germans. Joyce's Paris relics figure with especial vividness in the exhibition; they have been arranged with "piety and skill" by those of his Paris friends who survive, led by the catalytic Mrs. Maria Jolas and Léon's Russian widow, Mme. Lucie Noel. As was recently so accurately remarked by the literary critic George Slocombe, "No writer since Shelley had more friends. Joyce was the most reticent literary genius of this century. Of all his years of exile, he lived longest and perhaps most happily in Paris. It was here that 'Ulysses' was first published in book form, here he wrote 'Finnegans Wake,' here he gained world celebrity."

This compact collection contains examples of all his printed works in all known editions and translations; examples of his notebooks and of his published and unpublished manuscripts, in his uncertain, cryptic script; most of the consequential things that have been written about him or to him (such as letters from André Gide and T. S. Eliot)—all these knowingly gathered by people who knew him and know one another—together with photographs of him and of them, and of his family and their dwelling places, plus a curious gallery of battered oil paintings of his Irish ancestors that he usually carried around with him on his European odysseys. Three courageous American women publishers have notable places in the photograph gallery. They are the Quaker Harriet Weaver, of the Egoist Press, whose generosity permitted Joyce to finish "Ulysses," in twenty years; Margaret Anderson, who was the first to print it—or sections of it—in her *Little Review* and was haled into court by John Sumner for obscenity; and Sylvia Beach, of Shakespeare & Co., who first published it as a book. The exhibition is to be open until Christmas, after which the major part of it, including the family portraits, will be put up for sale—to an American university, many Americans here hope. It is the Complete Joyce, which should be kept intact as a reference library for time to come.

There has lately been a dog show, in midtown Paris; a cat show, becomingly held in the old-fashioned gilt purlieus of the Continental Hotel; a canary show, trilling in the lake pavilions of the Buttes Chaumont; an inexplicable boa constrictor, which was discovered comatose in the gutter of the Quai Voltaire beside the second-hand bookstalls; and General de Gaulle, stentoriously addressing a press conference in the Quai d'Orsay. He literally distinguished himself by saying loudly, "I am de Gaulle," then added the identification that he was the non-politician who was summoning the French to a national renovation—to be led by him, apparently, but only as a savior. He ended by saying tactlessly to the journalists, "I do not know how extensively your objectivity can deploy itself." His eighteenth-century Versailles court style of French and its vocabulary seemed more elegant, mordant, and unaccommodating than ever before. He accused the Anglo-Saxons of having recently fed Germany nothing but an apple of discord and said that a basic settlement in Europe could be obtained only by setting up cultural and economic relations between the French and Germans, the two dominant European nations, without interference such as there had lately been from Washington and London. As for the North Atlantic Pact, he thought that in case of trouble, American aid for France would eventually arrive, "but we would be dead by then, and such aid interests us relatively little." He favored the proposed change in the voting system which is agitating Parliamentary circles at the moment and by which the Communists would lose, according to their screams, a third of their representation—just how is not clear. But with voting changes would come a new election, which no political party except de Gaulle's craves. The betting is that he would steal a great many of the Popular Republican (M.R.P.) voters.

December 8

Paris has no bobby-soxers. Its popular *chansonniers* have to satisfy people who demand a more mature version of life and love than crooning and mooning. Yves Montand, who was trained by Edith Piaf, has now retrained himself and has been delighting adult crowds at the old vaudeville house L'Etoile, where he follows the acrobats and constitutes the entire last half of the bill. A former factory hand, he wears a kind of overalls on the stage. He has learned to make, in the best sense, a perfect show of himself—of his virile, suburban personality, his gift for miming and deftly timed gestures, his carefully clumsy dancing, and his happy hedonist's voice. His songs are short stories of sentiment and character. The favorites are "Clémentine," which is about a best girl; "Paris," about a lonely country boy newly arrived in the city; and "Les Feuilles Mortes," about lovers whose spring has changed to dead leaves.

December 28

By peering around, by glancing back, by peeking forward, Parisians already have a notion of what the new year of 1950 looks like, as the half-century date comes into focus. Some of the significant features, large and small, that make up the present approximation of the immediate future would seem to be the following. No. 1, Politics: The French government may fall. It is a question of passing a budget, a question of money in general, a question of taxes in particular. In this continuing crisis, if as many as three hundred and eleven deputies could agree long enough to vote their lack of confidence in the government, the Chamber of Deputies would also fall. France would be right back where she has been several times since the Liberation—back at the polls and a general election, which could do the Fourth Republic little good. No. 2, Good Cheer: The French called this the first prewar Christmas since the war. It was the first Christmas without rationed foods, except for coffee, and that was available in large quantities, as usual, on the black market. It was the first Christmas when there was every kind of merchandise that the heart could desire. Mostly, the French bought fine food for their stomachs. The poorer ones didn't spend much money except with the butcher and baker, dedicating their holiday purse to Maman's art in the kitchen and a gala dinner at home. No. 3, Forgiveness of Sins: A bill providing amnesty to collaborators with the Germans—if neither murder nor any other crime was involved—has been introduced in Parliament. If passed, this will make a very happy 1950 for eight thousand alleged collaborationists still in jail. No. 4, Music: For the past month, the Opéra and the Opéra Comique have been closed. The orchestras are striking for as much money as they would get if they played in a musical-comedy theatre, where musicians are paid a third more for playing less well. As state employees, musicians in the two Opéra orchestras can be given raises only by the Minister of Finance, who is too worried over the national budget to receive flute players and the like. No. 5, Iron Curtain News: Latest echoes from the Wroclaw trial have it that the guilty were, according to their lawyers, "led astray" by cosmopolitanism, by Sartrism (i.e., the ism of Jean-Paul Sartre), by American literature (no writer named), by Wall Street money, by Bevin, by Truman, by de Gaulle, and by Coca-Cola. No. 6, Intelligentsia: A book is about to be published here, as well as in the United States, called "The God that Failed." It is a collection of articles written by André Gide, Stephen Spender, Louis Fischer, Richard Wright, Arthur Koestler, and Ignazio Silone, who give the reasons for their disillusionment with the Communist Party—at one time they were all either Party members or fellow-travellers. No. 7, E.C.A.: This new year sees the beginning of the end of Marshall Plan aid, as originally timetabled. The E.C.A. and the possibility that disgruntled Washington legislators may close it down completely sooner than scheduled are far and away the most prevalent topics for editorializing in the Paris press, and such a move would drastically change the face of the New Year. For the most part, the editorials take the form of thanks mixed with questions. The thanks are for what America has given. The main question is: When is America going to start to receive? She has given Europe her money and her goods. When will she begin buying European goods? And what are Europeans supposed to buy from America with their shrunken money? A dollar deficit sometimes seems to be the only thing that the potential United States of Europe so far have in common. French editorial writers largely agree, in effect, that the Marshall Plan dollars were at first a warm, fertilizing wind from the west, causing postwar business to sprout again in the war-caked area of Western Europe. What they think is urgently needed now is a great wind that will blow in both directions across the ocean—and blow good and hard, like a splendid gale, for the rest of the twentieth century.

Jean Genet's "Journal du Voleur" has recently been published for the first time in a relatively cheap edition—that is to say, at nearly triple the price of the average novel. Ever since he was first published, in 1943, his works have mostly been privately printed and de luxe, enjoying those lavish publishing favors that make a book not so much forbidden to all as too costly for most. This new edition of one of his major writings puts him within range of general criticism. Genet was discovered during the war by Cocteau. By Genet's account, he was born illegitimate, raised in an orphan asylum, trained in a reform school, and became a thief. The commentary value of his "Diary of the Thief," a novel that is certainly unique in French literature, is its unsalvational viewpoint. He believes that society is not a mixture of good and evil but is a field with a fence across it—on one side are the evil, the outcasts, the misbegotten or desperate poor who for centuries have made criminality their special form of civilization, frame of survival, and group education. To his special damned land, to his country of wretched city alleys, abandoned only during protracted visits to prison, Genet gives his loyalty, his imagination, his fervid romanticism, and his talent as a remarkable novelist and writer of the French language; he creates, indeed, a sort of chauvinism, or patriotism, that until now has precluded an affection for, or even contact with, any other scene, as if he were a regional writer to whom no more interesting place exists on earth. His books, peopled by jailbirds, pimps, traitors, prostitutes, and vicious youths on whom, like a street light, the bright glare of his lubricity suddenly shines, are *sui generis*. They are nothing like the noted eighteenth-century picaresque writings of Restif de la Bretonne, nor are they remotely similar to the calculated tableaux of the Marquis de Sade, which were padded out, between the acts, by philosophizing. Genet writes of criminality and wickedness as naturally as Conrad wrote of the sea or as Hardy wrote of the landscapes of Wessex. Undoubtedly the least sought after of Genet's works is his brochure "L'Enfant Criminel," published a few months ago. It is a brief description of the brutalities of the French reform schools he knew when he was an adolescent, and it was written to be read by him on a national radio program called "Carte Blanche," accepted, and then turned down. In contemporary French letters, Genet is the lone *fleur du mal*.

1950

January 11

The old *Nouvelles Littéraires,* which, if only from pleasant habit, is still everywhere in France the most cherished of the Paris literary weeklies, brought out a stately 1900-50 number at the beginning of the year, to commemorate what important men have been thinking, creating, writing about, or appreciating as readers during the past half century. This special issue's focal point was the tabulation of a poll taken by the editor among two hundred notable *personnalités françaises,* who were asked to select Les Dix Phares (The Ten Lighthouses), or the ten brains that gave the greatest amount of light in that period. Einstein, one of three foreigners selected, won the place of honor at the top of the list, with the approval of seventy per cent of the voters. Behind him came Bergson, Proust, Debussy, Gide, Valéry, Prince Louis de Broglie, Freud, Picasso, and, last, the poet-diplomat-playwright Claudel, with a score of fifty-two per cent.

Some of the illuminators chosen by a smaller number of the voters were Hemingway ("to represent the American influence on French literature"), General Eisenhower, Marshal Foch, Rilke, Renoir, Rodin, César Franck, and Fleming—the discoverer of penicillin. By all odds the queerest light under a bushel basket, designated by only one admirer, was the witty, deaf, poison-pen, anti-Semitic, anti-republican, pro-Fascist, royalist old Charles Maurras, of the defunct *Action Française,* now finishing his life in a Lyons prison as a Nazi collaborator.

Of the many letters of explanation printed along with the nominations, one of the most interesting was from Maurice Schumann, Popular Republican Deputy and skillful, cultivated French politician. His contribution to the *Nouvelles* attained a cultural, philosophic, and literary tone that could well make the average Washington congressman's eyes bulge. "When I received your inquiry," Schumann wrote to the editor, "I began to think out loud, and the result is the following. My choice is Péguy, Bergson, Alain, Debussy, Picasso, Claudel, Maritain, Mauriac, [René] Grousset, and Mme. Curie. I am in no way certain that these ten are the most characteristic and important of the terminating half century, but I feel and I know that they built my universe. The only lecture by Mme. Curie that I was privileged to attend, even though I was incapable of comprehending it, extended my horizon forever after. Thirty-six years ago, the real Picasso already showed that he knew how to draw like Ingres. Mauriac had almost finished his famous novel 'L'Enfant Chargeé de Chaînes.' With surprise, I note that by instinct I have excluded from my list Barrès, Gide, Valéry, and even Proust. Did they enlighten me or lead me astray? I shall not know until my children are mature in their turn. All this merely confirms my idea that today the men of my generation are essentially 'survivalists,' a fact that condemns them to tragedy but confers upon them a singular dignity."

Coca-Cola has finally reached the diplomatic level here. According to the French newspapers, the United States Ambassador, David Bruce, in line with duty, recently demanded an audience of Premier Bidault to talk Coca-Cola. (He had already consulted Finance Minister Petsche on the same refreshing subject.) Last year, consideration of what the Communists gloomily called the "Coca-Colonization of France"—the projected distribution of the beverage on an enormous scale—was discussed by the French Cabinet and was mentioned, usually without affection, at several meetings of Parliamentary committees. This year, the Communists have a brand-new damnation of our national drink. They declare that the Coca-Cola distribution organization will double as a spy network. This harrowing conclusion is based on the fact that the head of the Coca-Cola bottling concession for North Africa, whose offices are in Casablanca, is Mr. Kenneth Pendar, who four years ago wrote a book called "Adventure in Diplomacy," which described his war work in that country as a vice-consul—work that included such unobjectionable tasks as making household arrangements for President Roosevelt and his party during their conference visit there. Today, the Communists see spies everywhere, and to them Coca-Cola is just plain Wall Street hemlock.

There has also been an equable but almost equally inhospitable editorial on Coca-Cola in *Le Monde*. After noting that American chewing-gum wrappers strew Paris boulevards, that our tractors plow French fields, that du Pont stockings sheathe Parisiennes' legs, that Frigidaires chill gourmets' foods, that anyhow Coca-Cola has been selling slightly in France ever since 1919, and that some new concessions for it will be assigned to, and thus will give profit to, French firms like Pernod and the Glacières de Paris, *Le Monde* penetratingly pointed out why many intelligent wine-drinking French shiver at the prospect of also drinking two hundred and forty million bottles of Coca-Cola a year, this, according to *Le Monde,* being Coca-Cola's official quota for France. "What the French criticize is less Coca-Cola than its orchestration," *Le Monde* ruminated, "is less the drink itself than the civilization, the style of life of which it is a sign and in a certain sense a symbol. For the implanting of Coca-Cola in a country is generally accompanied by advertising in the American manner, with red delivery trucks promenading publicity, neon lights, and walls covered with signs, placards, and advertisements." It should be added that an American Coca-Cola official here has since declared to the Paris *Herald Tribune,* "We don't plan to hang any red signs on Notre-Dame or the Eiffel Tower," the second of which, as a matter of fact, used to bear a gigantic, blinking red advertising sign that spelled "Citroën." "Certain very sound citizens," *Le Monde* continued sadly and wisely, "consider that, since many French customs have already regrettably disappeared, those that still persist should be defended. For it is now a question of the whole panorama and morale of French civilization."

Paris has sartorially revolutionized its traffic police and has turned traffic itself upside down in an effort to speed the circulation of automobiles through a city in which some of the streets have always been too narrow—even for a couple of the King's cavalrymen abreast. To heighten the traffic policemen's visibility at night, on the main avenues they now wear elegant white capes, white gauntlets, and a heavy white band around their kepis, which make them both chic and noticeable. As an experiment, a few of those on the Place de la Concorde have also been equipped with what looks at night like a magic wand—a glass baton, powered by an electric battery, that turns white for "Go" and red for "Stop." Since most motorists stop anyway to get a good look, the practice so far has been unfair to the theory. The shakeup in the traffic rules has given Paris twenty-nine new one-way streets, including the Rue de Rivoli, and a few more precious parking spots. In general, the system has been easy to follow. It is mostly the exact contrary, as to directions and turnings, of what it was before. On the day the new system went into effect, the Place de la Concorde, which was the most radically affected, was transformed into a motorized merry-go-round about the Egyptian Obelisk (thirteenth century B.C., period of Ramses II). The uproar of horns was so tremendous that pedestrians lined the square to jeer at and, for once, enjoy them.

January 25

"Le Deuxième Sexe," a long, thoughtful work by Mme. Simone de Beauvoir on the peculiar, even awkward place in civilization that first nature and then man have put the human female, was recently published here. It will eventually be published in the United States, where it should interest readers of the Kinsey Report. The American male would surely be cheered to read of his sex's age-old, uninterrupted, systematized position of superiority over the little woman, though, naturally, he would have to leave her and his home and come to Europe to live to get much out of the situation now. Some of the book might seem archaic to American women (whom the author is not writing about), because the difference between the average American male and the average American female has been reduced to a minimum by our un-European way of life. Mme. de Beauvoir is discussing the maximum, as she finds it in France. French publishing figures show that French readers are not interested in reading books about sex, which hers basically is. The Kinsey Report in French hasn't sold at all. Freud's works were influential but never popular. Mme. de Beauvoir's book has received some rough, unjust reviews. Yet it is a serious, provocative study, unlike any ever produced by a French *femme de lettres* before.

Mme. de Beauvoir has been a philosophy teacher and is a novelist and an editor, and in "The Second Sex" she functions as all three. Her volume is carefully documented, though perhaps there are too many case histories of frigid and unhappy women and not enough of contented women. Its quotations from world literature, including novels—there are few from science—are numerous and scholarly. The book is a dry, wry job, written straightforwardly by an intelligent, determined woman, a practiced writer, and a watchful observer who, having inherited three thousand years of European civilization, has something conclusive on her mind that she wants to say. Her opening premise, which would certainly not be everybody's, is: "One is not born a woman; one becomes it. No biological, psychic, or economic destiny [makes woman] but the ensemble of civilization, which elaborates this intermediate product, between the male and the capon, that is called feminine." More mildly, she thinks that woman's defects and limitations are precisely what society has desired of her and has perfected in her; that her character, such as it is, is the result of her situation; that sex has been hard sledding for most females, and that this is the fault of man and nature; that "the eternal feminine" is hocus-pocus and a male alibi; and that the battle of the sexes should be not denied but explained. A long philoanalytical chapter entitled "Justifications" was to this *deuxième-sexe* reader the most interesting part of the book.

Not in a quarter century have the food markets of Paris been fuller or more tempting. In the Rue du Faubourg-St.-Denis, there is a two-hundred-yard stretch of food shops and street barrows. The hucksters shout their prices and the shop apprentices have become barkers, standing at the shop doors and bawling about the luscious wares inside. In the *charcuteries* there is a mosaic of every known dainty—turkey pâté, truffled pigs' trotters, chicken in half mourning, whole goose livers, boar's-snout jelly, and fresh truffles in their fragile bronze husks. In the poultry shops, there are Strasbourg geese and Muscovy ducks. In the entrail shops, there are indescribable inner items and blood sausages. At the fish stalls, there are costly deep-sea oysters and enormous, hairy sea spiders, to be buried in mayonnaise. The street barrows are filled with bearded leeks and potential salads. The Rue du Faubourg-St.-Denis is not a rich district of the city, but these days it offers a Lucullan supply. Food is still what Parisians buy if they can. It is a nervous means of getting satisfaction, a holdover from the lean years of the Occupation. As for other commodities, also here in abundance, people say they cannot afford them. Prices in the annual January white sales have dropped, and in men's shops shorts, pajamas, and shirts have been marked down from twenty to sixty per cent, but no one is buying. Bankers sourly declare that only a conventional postwar depression will jostle prices and things and people down to normal again, at which point people will buy an undershirt and lay it away, instead of buying pâté de foie gras and eating it.

May 3

There has been considerable literary talk about, and certainly substantial spiritual refreshment in, the revival of Church faith and practices, which started during the sadness of the Occupation, as if postwar French life offered almost the same melancholy and material discouragement as the war years. Among the major contributors to the reanimated Catholic faith are the posthumous books of a Parisian Jewess, Mlle. Simone Weil, who was converted to a personal, perhaps unorthodox Catholicism. Her first book, which was printed two years back and has since gone through several editions, was called "La Pesanteur et la Grâce" ("Weight and Grace"), a title that expresses what was her belief—that the weight of original sin is the law of creation and divine grace consists of man's de-creating himself and losing his ego. Her second book, "L'Enracinement" ("The Roots of the Matter"), was published last year, and the third, "L'Attente de Dieu" ("Waiting for God"), is now on the press. The Weil books have touched the lay-religious soul more than any other personal confessional since Amiel's "Journal," that celibate revelation of Swiss ethics that came out of the last century. The Weil offerings are more apostolic and Pentecostal, though Mlle. Weil's gift of tongues, if fiery, was hasty and not literary. Sure of an early death from semi-starvation and tuberculosis, she did not bother to leave her report of her "inner purification" in any more polished form than a series of notebooks. She was a Parisian, a pupil of Alain, was a University of Paris graduate in philosophy, fantastically erudite in classical languages, and a schoolteacher outside Paris until the Occupation race laws drove her to the Unoccupied Zone; was non-Communist but militant Left; had worked in the Renault factory to detach herself from her intellectualities; and had labored in the fields, where she insisted on living solely on wild blackberries. She followed her parents to New York, then soon afterward went to

London to broadcast for the French Resistance, there refused to eat more than her French friends in France received on their meagre ration cards, and there dwindled away, aged thirty-four, in 1943.

Mlle. Weil endlessly talked mysticism in a monotonous voice, and suffered almost uninterruptedly from headaches. She became so Christian that she was once in trouble with the Church for resenting the inclusion of the Jewish Old Testament in the Bible. To the general reader, her first book seems to be her best, and its oddest, most affecting chapter is the one called "Le Moi" ("The Ego"). Her illuminated idea was that the "I," the personality, the ego, must be shed and effaced. (There is also a thought-provoking chapter entitled "Effacement.") She says many strange things that the psychoanalysts, whom she scorned, have rarely said so well, such as: "We really possess nothing on this earth except the power to say 'I.' This is what we must give over to God—that is to say, destroy. Absolutely no other free act is permitted us. Though nothing should be gained from victory, one supports death for a victorious, not for a defeated, cause. . . . If you help unhappy people and are ill-treated afterward, you are merely carrying a frail part of their unhappiness. . . . Pharisees were people who relied on their own force to be virtuous."

Business is bad in Paris. Owners make profits, all right, but unexpected new jumps in the prices of materials whittle their profits down. When workingmen win their strikes for higher wages, the increase goes into the next new price, and in the end everybody is a consumer, carrying an increasing burden. It will be an aesthetic blow for Paris itself and a tragedy for the myriad workers involved with the *haute couture* if the big dressmakers do not pull out of their current financial doldrums. Skirts are short again, but dressmakers' faces are as long as trains.

Spring has been like leftover winter here, with rain, sleet, and winds until May Day, when the traditional wild lilies of the valley, gathered from neighboring woods, were green, scentless, and very expensive on their stems. But in the forests the chill rain and clouds finally produced some compensating joys. For those willing to trudge among the trees to find them, there has been a wondrous mannalike crop of morel mushrooms, which look like flaccid sponges, are the gourmet's delight, and are considered the finest table gift hereabouts. And the neighborhood nightingales, which had been mute under April tempests, burst, in relief, into great choruses of churring, trilling, and musical pining under the welcome full moon, white and virile, with its man's face showing extra plainly, that brought in the first nights of May.

May 18

Foreign Minister Robert Schuman's surprise project to pool France's and Germany's steel and coal production has so enraged the French coal and steel magnates and the French and German Communists that they are all fit to be tied—for once, in the same bundle. In one way, at least, the Schuman project has pleased many Frenchmen, even those who, owing to caution, hatred, or memory, think Germany should always be pushed down and never helped up so that she can stand again on her own, and probably France's, feet. The average Frenchman is pleased because he believes the Schuman proposal of a Franco-German pool is the most creative of all the postwar ideas so far, second in importance only to the Marshall Plan and superior to it in realism, since Schuman's proposal literally means business, whereas Marshall's kind notion is regarded as meaning charity. Psychologically, too, the Schuman project has given local spirits a lift; it has been inspiring for Frenchmen to see their defeated France, once mistress of Europe, actually taking a positive, unwobbling step forward on her worn old high heels, without an initial friendly push from England, more recently the balancer of Europe—a push that France has waited for since the war's end. Indeed, many French feel that Schuman has at last furnished an answer to France's most worrisome peace problem: whether France should try to make a Europe without England or not try to make a Europe because of England?

Of course, many French also think that the Schuman plan may just possibly provide the answer to the even older German postwar problem, for it obviously gives new hope to Western Germans, already impatient after five years of fairly benevolent, bumbling, and costly Allied occupation, so unlike the Germans' own years of cruel, well-planned, locustlike occupation of all Europe, from Poland to the Atlantic and the Mediterranean. In the past hundred and fifty years, France has tried everything with the Germans except helping them, which to the French has never seemed natural; she has conquered them and been conquered by them, has paid them and attempted to get paid by them, has been occupied by them and has occupied them, has been overharsh with them and contemptuously loose with them. Schuman's shrewd plan is regarded by some people as the first intelligent French effort to control modern Germany's vigorous industrial genius, which is always working on what look like baby carriages but turn out, when put together, to be cannons. No people in France, or anywhere else in Western Europe, are as hysterical about the possibility of a war with Russia as the Americans, or at least those Americans on holiday here, whose Fifth Avenue or Golden Gate would probably be the last to face hand-to-hand fighting with Russian soldiers, if such a muscular anachronism is conceivable in any next, scientists' war. But the French would like to feel that if there is, God save us, to be another war, the German cannons will not this time be pointing in France's direction. To the French, the Schuman plan seems a bid for business, salvation, consideration, and even a protracted peace.

June 1

Right now, Paris is a refined boom town. For its natives, there is too much to see and hear—a blend of intellectual entertainment and honkytonk. Hotels are jammed to the bathtubs with tourists from all latitudes, with enough cash for everything, from Furtwängler's magnificent and high-priced concerts (Berlin conductors and orchestras are ever the dearest in Paris: *aber warum?*) to the Folies-Bergère and its chocolate-tinted American Peters Sisters. In the latest European version of themselves, they descend through the theatre's ceiling as crooning, overweight parachutists. At the Opéra, also at extra-high prices, Flagstad has been heard in "Götterdämmerung," and Max Lorenz merely seen, except in the last act, for which he saves his voice. One modest item in this vast grab bag has been the exquisite performance, at the suburban Château de Sceaux, of Lully's "Idylle sur la Paix," with a libretto by Racine. This was only its third presentation since its première in 1685. Most of us went as far as we could by subway, then trekked a few hearty miles through Le Nôtre's lovely topiary gardens, with their pollarded linden trees imitating arches and walls. In their day, Louis XIV, for whom the idyll was written, and his guests arrived in greater comfort, in gilded coaches. He had momentarily beaten all his Spanish, Austrian, and English royal neighbors, including his relatives, and had burned Heidelberg University, and France, in its apotheosis, was starting a short breather of peace within his fifty years of almost continuous war. So there, last week, sat a few hundred of us in the château's battered orangerie while singers chanted to us, as once to him, "Let him reign!," which has too often seemed the theme song of this Continent, now including Russia. This highly enjoyable Lully-Sceaux concert of ancient music was produced and broadcast by the national radio broadcasting units.

This is the season for France's political parties to clean house in the privacy of their annual caucuses, to discover what they think about their rivals and themselves, as well as about what they have done during the Parliamentary winter and what they had best hustle and do about pending bills before the summer vacation. The Anarchists' Federation (which is apparently merely Trotskyite), Bidault's M.R.P., and the Socialists have already held their official huddles. The forty-second annual Socialist Congress, which met at the dingy Left Bank Palais de la Mutualité, was the most important, because, as the entire Paris press commented, it was the first in fifty years without the guiding presence of its late perennial leader Léon Blum, and its duty was to find a new leader. "No one can succeed Blum except the entire party itself!" cried one zealous delegate on the opening day. "Everyone has confusedly felt this. It was good to have it so stated," gravely editorialized the Party's *Populaire* next morning. The Party, noted usually for its many noisy divisions of conscience, actually integrated itself to a rare degree. For one thing, though bitterly anti-Communist, it demanded, with what it regards as realism, a Chinese Communist delegate to the United Nations. It voted to uphold the Bidault government but to continue disdaining, as it has done since February, to take part in it. This was a sort of blow to the government, but one that it will probably weather. Anyhow, the real crisis this spring for the government, as well as the parties, is the approaching electoral-reform bill, which advocates a complex new proportional-representation method of voting. This reform, for reasons that are also complex, would hit Communist candidates the hardest but could also upset every party's calculations.

At the moment, Communist propaganda is enjoying the most extraordinary success, especially among non-Communists, that it has ever had in France. The Party's Partisans of Peace, whose first meeting was held here only a year ago, is now its most powerful official spearhead against the Atlantic Pact and its implementation, and it is also the greatest magnet for attracting international millions of men of all beliefs, all frightened of atomic war. The Partisans of Peace recently met in Stockholm, where it formulated the snowballing "Stockholm Appeal for the Absolute Interdiction of Atomic Weapons." Point No. 1 demands "the establishment of rigorous international control [or just what America long ago asked] to assure application of this interdiction." The third, and last, point is "We call on all men of good will in the world to sign this appeal." The appeal was widely reprinted here, with a coupon providing space for a signature and the signer's address, and gives a Party address to send it to. The Stockholm Communist appeal has already been signed by a hardly credible galaxy of prominent anti-Communist Frenchmen. This past week's signatures alone, received and headlined by *Les Lettres Françaises,* the Communist literary gazette, included such eminent public figures as André Mornet, former Attorney General of the Republic; Paul Mongibeaux, president of what corresponds in France to our Supreme Court, who was the judge at Pétain's trial; the two presidents of Parliament; illustrious members of the Goncourt and French Academies; and a mass of university professors, popular writers, star actors, and famous artists, including Picasso, unique in the week's list of signers for being already known for his Leftist sympathies. Fifty per cent of the population of Calais have signed, some French towns signed en bloc, eight thousand signed in Casablanca, nine Swiss communal councils signed in a body, nearly five thousand British civil servants signed, forty thousand Chinese signatures came in in fifteen days (surprising only for the speed), and still according to *Les Lettres Françaises,* sixteen thousand *"New-Yorkais"* signed in Madison Square Garden. The new Stockholm appeal is really old; it was an American appeal when it was first launched—and refused by the Soviets. Now it is offered by the Soviet Communists everywhere, who thus brilliantly impound the credit. Masses of cautious Frenchmen evidently regard it as Russia's first offer to consider a commission of atomic control, and as men of good will they have eagerly signed it as a kind of contract. Or so they hope.

June 21

France was shaken by the distant, duplicated tragedy in the Persian Gulf that recently befell two Saigon-Paris planes. They perished off the shore of the same remote island of Bahrein on June 13th and June 15th. As one commentator said, "In this double rendezvous with death, at a fixed hour, at a selected point, there is something that chills the blood. Something in man protests. He can accept the accident. But the coincidence terrifies him." The Left newspapers had something more positive to ruminate about. By chance, the sad news from Bahrein was synchronized with other information from the East, which the Socialist *Populaire* had just received. While everyone was wondering what was suddenly wrong with the Saigon-Paris run, *Le Populaire* told what at least was wrong with Saigon; a feature story headlined Saigon as "THE CHICAGO OF THE FAR EAST" and described it as complete with lupanarian dance halls and dens of vice and a Hotel Continental whose French boss had apparently just celebrated with a champagne dinner his second billion francs of profit. According to *Le Populaire,* Saigon's glory is, however, a French gambling casino called Le Grand Monde, which is reported to run the local gold and piastre traffic. *Le Populaire,* which is violently anti-Communist, also related that to avoid being bombed in France's Indo-Chinese war by the Communist Ho Chi Minh, Le Grand Monde pays him a million seven hundred thousand francs daily—on top of the five and a half million francs' casino-privilege money it daily pays the French government. Similar scandalous news, given earlier to *Franc-Tireur* by its Saigon reporter, François-Jean Armorin, led to his being beaten up and his life's being threatened by the vice ring, or so his last published reports said. There can be no question of further danger for him now. He was killed in the second air crash, en route to Paris with more news.

The Big Six are now meeting here to discuss the Schuman-Monnet Franco-German coal-and-steel pool—really a Big Six minus one, owing to the withdrawal of the British Labour Party from participation in the project. Ten of the Western Europe Socialist Parties are for it, leaving the British as the old-fashioned dissenters. Most French citizens feel that the Schuman Plan, if made to work, could establish new Franco-German relations, a new spirit of Western union, and, lastly, a new French prestige, which is what they usually mention first.

July 6

For the sixth time in two and a half years, the Socialists have toppled a French government. Everybody is understandably cross with them except the thousands of ill-paid state and civil employees, who make up the greater part of the Socialist Party and whose proposed salary increase is what the Socialist deputies are standing pat on as their price for Parliamentary coöperation. Whatever government turns up next will have some prickly problems. It must balance the budget and then unbalance it enough to find some loose cash for these underpaid functionaries and for France's disgracefully underpensioned war cripples.

Delegations of war cripples from the First and Second World Wars have just gathered from all over France and, after a tragic, hobbling parade down the Rue Montmartre, held a strange seven-hour bivouac, ending at midnight, on the Place de l'Opéra, with tents, campfires, chow, card playing, songs, reminiscences, and, of course, attendant ambulances. Their slogan was *"Ajustez Nos Pensions!"* A quick twenty-five-per-cent increase was the adjustment they had in mind. The *mutilés* claim that, what with the high cost of living and the government's delay in classifying them according to its complex system, old soldiers are being paid sixty per cent less than they were getting right after the First World War. All the cripples wore their medals, and many wore blue poilu helmets from World War I. A few were the faceless men, or *gueules-cassées,* wounded before the days of widespread plastic surgery. Some had rolled up a World War II khaki trouser leg to show a metal limb; others, blind or legless, were in basket chairs pushed by their wives. A straw effigy representing L'Argentier, the sym' ol of all Ministers of Finance, was burned in front of the Opéra steps. At midnight, people coming out of the boulevard cinemas met the bemedalled cripples, straggling, limping, or being wheeled back to wherever they came from.

The Grande Nuit de Paris fireworks on the Tour Eiffel, which served the whole city as a fabulous chandelier, were the finest ever seen here. Hundreds of thousands of people packed the bridges and squares from as far down at Notre-Dame to as high up as the hills of St.-Cloud to watch them. The *feux d'artifice* were created by the famous French pyrotechnician Maître Ruggieri, who has long gone without a first name and who puts on the annual July 14th fireworks in Paris and recently staged displays at the Montevideo Carnival, at Haiti's centenary celebration, and at the Monte Carlo investiture of Prince Rainier III. Shrouded in smoke, the Maestro personally operated his Tour Eiffel display from the second balcony, employing a twelve-button panel that set off the eighteen hundred component parts of the varicolored conflagration. Until eleven-thirty, the Tour, covered by a patina of floodlights, looked as if it had been freshly painted white. Then it was suddenly darkened, and from all three of its balconies burst an apotheosis of fireworks in showers, balls, streamers, rockets, and what looked like silver and gold clouds of burning snowflakes. To the handful of us gaping near the base of the Tour—oddly, the only empty space—the finale was a magical, thunderous bombardment of explosives, going off in artfully patterned flames, and above them the Tour seemed to rock peacefully in the night sky.

July 19

From France's vast fund of experience with wars—Western and Eastern, won and lost—the French people, out of friendliness toward us and worry over the grim Korean events, have been pouring advice into the ears of Americans here, and into their own newspapers. Paris gentlemen whom one hardly knows talk to one like Dutch uncles and military experts combined. Paris newspapers that have loyally supported their government's and their Army's unsuccessful policies in the Communist Ho Chi Minh-Chief of State Bao Dai war in Indo-China—now in its fourth year, and thus serviceable to them as a presumed prototype for our Korean war—have been particularly explicit in telling us what our errors are. And, indeed, errors are what they should be expert on. Paris café strategists have begun explaining what is really going on at the Korean front. Last week's most bruited corner-*bistro* rumor was that the American Army, as a brilliant ruse, would let itself be pushed from Korea, so that Japanese soldiers could take its place, in what would then become a kind of proxy war between mere natives, with

Uncle Sam and Stalin white-man spectators. This gathered such momentum that the Japanese Premier, Shigeru Yoshida, was forced to declare that his country can go to no war right now, since it is still occupied as a result of the last one.

In an editorial entitled "Sodome et Gomorrhe," *Le Monde,* which always regards the United States as insensate, like Lot's wife in her last phase, pointed out "Europe's increasing fear on seeing the result of American intervention in Korea." "Anguished Europe," it went on, "asks itself if it must again undergo the infernal cycle of invasion-occupation-landing-liberation, worsened by the presence of the Russians, who would be far more terrible than the Germans were. Furthermore, Europe is alarmed at the revelation of American military weakness. Europe placed all her security on American arms, neglecting," *Le Monde* admitted angrily, "to build up her own, only to perceive that there is no American Army and that it takes only a few North Koreans to put in check the nation that Europe thought the most powerful on earth." *Le Monde* added, perhaps for our comfort, that the grave lesson of Korea would doubtless bring about repercussions in Washington similar to those that followed Pearl Harbor.

The Academician François Mauriac, in an editorial in *Figaro,* took the most transcendental tone toward Americans: "Should I say all that I think? The American people must revise their whole concept of life if they mean to save human liberty. A people cannot be at the same time the most happy and the most powerful in the world. Between the power of the nation and the happiness of individuals, Russia, for her part, has made her choice." Americans, "on whom weighs the heaviest responsibility that any people has ever assumed," are, Mauriac said, learning the harsh lesson that any great policy, and especially one of such grandeur as saving the world, demands arms behind it. He took issue with Admiral Nimitz's statement that the early, rapid victories go to an aggressor and the final, slow victories to a democracy, and contended that "on the contrary, Soviet Russia shows no haste. She advances a pawn on the chessboard after taking her time, which can be weeks, months, or years, and then watches the reactions of her adversary, which are doubtless those she expected. A man of medium intelligence," he added helpfully, "can obtain an idea of her possible moves by looking at a map of the world."

For centuries, the French have been experts on the subject of prestige. They consider that if America is pushed out of Korea—and French military men say that they believe this can be avoided only by a *contre-offensive de grand style,* for which we lack the necessary men and material—the lowering of our standing in the Orient will change the pressure-politics map of Europe so drastically that no one can precisely envision what that change would be. As prestige experts, the French are more alarmed at what is happening to us in Korea than they are at what has happened to them in Indo-China.

November 2

The first hoar-frost has finally blackened the flowers in the Tuileries Gardens, where a palette of extended French autumn was lying undisturbed. Weather prophets have warned that the Seine will likely freeze over before the year is out. The winter of 1950 has come in cold but peaceful—in Europe, at least—after a summer of sultry fears of another world war. The harvests have been good in the French countryside. With a feeling of relief, Paris has started its winter routine. The weekly ballet nights at the Opéra have begun again, and the Colonne, Lamoureux, Pasdeloup, and Conservatoire Orchestras have tuned up for their weekly musical rivalry. The Comédie-Française, apparently with one eye on the thermometer, is opening its season with a new production of "Le Conte d'Hiver"—Shakespeare's "The Winter's Tale." This is definitely a revival. It was first played here, in French, in 1631. The reconvened Parliament, occasionally the most interesting public spectacle in town, has just held its first all-day-and-all-night sessions, one of them signally historic and all of them marked by that violent disparity of views and casual absenteeism of members that seem normal to democratic chambers everywhere today. Frequently, there were more visitors in the not too commodious galleries than there were deputies on their benches, though a government of France was turning her history topsy-turvy by voting that Germany should be rearmed.

Until the concrete, complicated rearmament scheme known as the Pleven Plan was adopted by Parliament, idealists here had hazily hoped that a federated Western Europe, that elusive twentieth-century goal, would somehow or other be founded on peace, balloting brotherhood, and fine phrases. The new French plan offers a United Western Europe founded on Ruhr coal and steel (the

Schuman Plan), plus a Western European army of nationals who, in accordance with a new ideal, would fight with instead of against one another. In one respect, René Pleven's plan rises magnificently above the memory of the old melees. His is the second French project for Germany in modern history—the first was the Schuman Plan—in which the French motivation has not been *"La Revanche,"* a patriotic slogan that was most popular after 1870, when French revenge seemed a pleasant possibility, and that has behind it a deeper French feeling that goes back to Waterloo, for which Blücher is still remembered and hated here, though Wellington has been practically pardoned and forgotten.

Obviously, many citizens, recalling what Hitler's labor battalions, theoretically trained with shovels, were able to do to them in 1940, are confounded in 1950 by the idea of France's handing any German even a popgun, in whatever federated wrappings. Also, the special Gallic satisfactions in the Pleven Plan, such as its ranking of the German soldiers as helots, without patriotic perquisites or their own military leaders, were badly shaken during the Parliamentary debates by news from West Germany that, aside from Chancellor Adenauer and the rich industrialists, the Germans did not want to be armed at all, *danke sehr.* Nevertheless, to the three-quarters of the French who are not Communists and who are, generally, realists, the Pleven Plan has a strong appeal. It seems a genuinely creative and hopeful diplomatic idea for a new relationship with Germany and for a modern Western Europe. The military risk seems realistically minimized by the Pleven Plan's being built on top of the Schuman Plan for the handling of the Ruhr's coal and steel, which the French also consider a creative and hopeful idea. For, as the French bitterly realize today, what modern European wars are prepared on and fought for is industry, or dialectic ideas on how to run industry. Despite the setback the Pleven Plan received in Washington, the French feel that with these two contributions France has become once more the planning brain of Western Europe, or as much of a planning brain as she can be with England so standoffish and America so powerful.

1951

February 20

The French, as you hear them talking now in Paris, seem certain that General Eisenhower and Premier Pleven, while visiting each other's lands a few weeks ago, presented, in turn, the most valuable gift that lies in the power of such simple, democratic plenipotentiaries. Each gave the gift of explanation. The French, on their various levels, profoundly appreciated the gifts of both, and are still inspecting and analyzing them. To what Eisenhower had to say, Parisians have paid their strangest, sincerest compliment; his brain, they say, did not smell military, even in this crisis in history. They feel that he spoke as a great and modest humanitarian. And the French even understand themselves better after what Pleven said about them in Washington. This is a treat for them. Nobody had properly explained them to themselves since the Great Debate began. The facts of life in a continuing crisis, or of death in an invasion, have changed little, but now these facts are being looked at as part of a possibly—just possibly—controllable destiny, rather than as mere fate. To the humble French, Europe still remains a matter of geography, not ideology. France cannot move off the map, away from the prospect of supplying some more of the most beautiful ruins, architectural and cultural, of modern history. Many middle-class French see Europe as faced with the choice between the Soviet threat of peace and the American threat of war. Many educated French see Europe as in danger of sliding from the sheer weight of history into a hegemony: a state of tight physical union such as was aimed at, through conquest, first by Napoleon and then by Hitler, and is now aimed at ideologically by the Soviet, a hegemony such as only the Union of Western Europe could have made—or may now belatedly make—painless, noble, and voluntary.

There is still a good deal of anti-Americanism here, brash or subtle. Many French bitterly accuse America of being whatever those particular French can be sure it is because of never having been there, including too young. It is those same French who weakly make excuses for elderly France, as if she were a handsome old lady of privilege who had gone off to live upstairs in furnished rooms, in a mansion that had once been hers and was still elegant in all the world's eyes and had all Europe as its topiary garden. To those disappointed French, France seems to await a history that will be supplied to her painfully by one of the two colossal foreigners bearing down on her from East and West. There is one statement that Americans here are becoming aware that they never hear. They never hear the French explain how it happens that France, with her superior old culture and knowledge of how Europe should be handled, finds herself in such a position that ignorant, amateur, bumbling Americans have been called in to do the job.

Eisenhower is reported to have said, smiling, that he wasn't surprised by the Communists' demonstration against him in front of the Hotel Astoria here but that he was surprised there was no counter-demonstration. According to the government order of that day, January 24th, "The Communists' proposed demonstration against him who was chief of the army of liberation for France is treachery and a scandal not to be tolerated." It added that any foreigner taking part (and the foreigner might have been an American who had come only to counter-cheer) would be expelled from France the next day; any participating French civil functionary of "permanent" grade would be disciplined and any such functionary of "temporary" grade would be dismissed; any employee of a nationalized industry would be suspended or fired. These are orders that now apply to any unlicensed demonstration in Paris. They are probably what kept thousands of French and foreigners from going to the Champs-Elysées to shout *"Hourra pour Aike!* Hurray for Ike!" in a melee in which twenty-five hundred Communists were arrested and five journalists were so beaten up by the police that the Chief of Police apologized.

When Pleven mentioned French "neutralism" during his visit to Washington, he put the word into a wider circulation than it had previously known. It was coined in Paris more than a year ago by the Academician Professor Etienne Gilson, of the Collège de France. Gilson is the most erudite, most venerable exponent of medieval Catholic philosophy, France's greatest authority on Thomas Aquinas, and possibly the greatest living Thomist. In a series of articles published in *Le Monde,* he first used the word *"neutralisme"* in his attack upon the Atlantic Pact. (French journalism now regards *Le Monde* as neutralism's editorial center.) Neutralists claim to be not defeatists but believers in the proposition that Europe is sick from many wars and another war would kill her, and that only America and Russia, rotting with wounds, would survive. Gilson recently went further. He resigned his Collège de France chair, stating that he would live in North America—in French Canada, by preference—rather than remain in France to face another war and another occupation. His decision and departure made a big stir in intellectual circles. In November, when he was at Notre Dame University to give a series of lectures, he made some comments about neutralism, which were later firmly opposed in an open letter by Professor Waldemar Gurian, of the same institution. This letter was printed over here in the weekly *Figaro Littéraire,* causing additional intellectual commotion. Although the daily newspaper *Figaro* is the center of Paris antineutralism, its editor, Pierre Brisson, who rarely editorializes, created a third climax in January with his brief anti-neutralist editorial headed "Les Insexués," The Asexuals. In conclusion, he harshly said, "The wild volition to avoid war makes no sense unless it is affirmed virilely, with readiness to make a supreme effort. Neutral France does not exist. There are neutral men, in the sense that they are *insexués.* They are few. It is not by them that we expect the country to be saved." Brisson doubtless speaks for more French than does the relayed voice of Thomas Aquinas.

According to intimates of André Gide's household in the Rue Vaneau, as the great writer lay there dying—and aware that he was dying—he said calmly, "If anyone asks me a question, make sure I am conscious before you let me reply." Thus, at the age of eighty-one, did he protect his genius for lucid responsibility as his final, most valued earthly possession.

Of those candid critical obituaries that are a sort of philosophic specialty of literate French journalism the morning after a great French figure dies, that of *Figaro's* devotedly Catholic editorialist

François Mauriac was judged perhaps the most masterly. Under the title "Un Destin," he wrote, in part, "Let us not gild his mortal remains. Let us not pretend to have misunderstood the redoubtable instruction of the immoralist. If what we Christians believe is true, today Gide knows what we all will soon know. What does he know? What does he see?" Mauriac asked these questions suddenly, evidently seeking at his editorial desk at midnight to determine precisely the knowledge and sights of immortality that Gide, who was born a Protestant and became an anti-Christian agnostic, had always derided. "He was the writer farthest from art for art's sake," Mauriac continued. "He was a man engaged in special combat for two causes, of which the more apparent was scandalous—to excuse, even to legitimatize, a certain kind of love. Worse, in his youth he broke with the moral law as taught by both Christian churches. No one could have staked his bet against Christianity with more sangfroid and reasonableness than Gide did. Few would have dared to overthrow Bossuet's tribunal of conscience, which condemns all crimes. This Gide accomplished with a tranquillity, a serenity, and a joy that were frightening. We must be living in a casual epoch indeed when his being awarded the Nobel Prize caused no stupefaction, no terror. His literary work? It remains among the most significant of our time. The Gidean critical spirit was incarnated in the *Nouvelle Revue Française,* which he helped to found. 'Les Nourritures Terrestres,' 'L'Immoraliste,' 'Si le Grain ne Meurt,' and his 'Journal' will long preserve as human leaven the ferment that it was his mission to provide. He followed his own path, wrapped in his famous long cape, with a constant solicitude for culture and dignity, with the noble tread of a grand seigneur of great race. Gide was not a poor sinner but a strange pilot."

Gide received the homage of Paris lying on an iron bed in his Rue Vaneau study, wearing one of the heavy sweaters but without the cape and the high, homemade cap—almost like a fez—that he used for many years to protect his chill flesh against the cold. He will be buried near his family's Normandy property at Cuverville, which he so lovingly described in "La Porte Etroite."

March 1

All that Parisians finally understood about Parliament's several mixed-up electoral-reform projects was that their government could fall among, and because of, them. French citizens are groaning in anticipating of March 15th, the last day for paying their income tax. "Ulcerated by the bigamy" of his Radical Socialists in making strange political bedfellows of de Gaullistes, the Radical Socialists' grand old man, Edouard Herriot, has, at the age of seventy-eight, chastely resigned the leadership of his Party. Everybody in Paris walked during this Monday's bus and Métro workers' strike, which was pretty close to total. Prices on practically everything have suddenly leaped alarmingly, including bus and Métro tickets. Everybody is worried and nobody talks of anything else. Compared to 1950 prices, knitting wool is up 65 per cent, workmen's blue jeans 25, kitchen pans 8, paper 60, shoes 15, coffee 55, ladies' hats 20, and pork chops 50. The weather has been dramatic, with tempests, wind, sun, and downpours all taking their turns within the hour, for several hours daily and day after day. Now all is blanketed down by fog. On the hucksters' flower carts on boulevard corners, the first vernal Parma violets are selling cheap and gay at four hundred francs the big bouquet—a fistful of scented purple that shades in with the city's lavender mists.

March 22

For nine days recently, France was without a government. In other political crises and stalemates here since the Second World War, Léon Blum was sought out for consultation, even by politicians to whom Blum's Socialism was anathema, for they admitted his qualities as counsellor, his skill at arbitration, his political experience, and, indeed, the value of his annoying ideals. In the past few weeks of confusion, he has been much missed, if only because he might have kept his party in line and prevented its members from saying one thing in Parliament and voting another—like everybody else. Until his death last year, Blum was the brains—and for much of his career he was the body as well—of contemporary French Socialism. He was more assailed in peace and longer imprisoned during the war than any other noted French leader. He would no doubt have been surprised at his obituaries in other countries, and even in his own, which mourned him as France's grand old man.

Blum died, aged seventy-seven, in his suburban cottage near

Jouy-en-Josas, a year ago, on March 30th. He was both a loved and a hated figure. He was known the world over as the great leader of modern French Socialism, and yet he was an unsuccessful politician. He was also the political figure who—because of his long life, his Jewish origin, his humanitarian aspirations, and his intellectual refinement; because of the schizoid conflict between his acute national patriotism and his devotion to international visions like Socialism itself, pacifism, and unrevolutionary proletarianism; and, above all, because of his prescience as a social reformer—personified the whole complicated, struggle-ridden, sad history of the French Third Republic.

Like Blum, the Third Republic began its life at the time of the Franco-Prussian War and the Commune, moved through the anti-Semitism and anti-libertarianism aroused by the Dreyfus Case (in which Blum and the rest of the liberal element in France fought and suffered), endured the first German world war of aggression (in which Blum's Socialist pacifism failed and France won), then rose to the peak of novelty, for both him and the state, when he became the first Socialist, first Jewish Premier of the very money-loving, very Catholic republic during its Front Populaire government. And, for both him and the state, the Third Republic ended, as it had begun, with an aggressive German war, during which, as a significant symbol of France's defeat, an old Blum, the perfect quadruple hostage—Socialist, liberty-loving intellectual, Frenchman, and Jew—was imprisoned in the Nazi concentration camp of Buchenwald.

Blum's character, interests, and talents made him, until he was forty-eight years old, the most unlikely future important politician then to be found in any fashionable literary salon in Paris. Like George Bernard Shaw, he was of the vintage of comfortably well-off ethical intellectuals who espoused Socialism as a means toward justice for all, and especially the working poor. Born in the drab Rue St.-Denis, Blum always cited its humbleness, as though it were a political birthright, to his working-class followers, with whom he certainly had no other earthly connection. His father's ribbon business early lifted the Blums into the prosperous bourgeoisie and its neighborhoods, and lifted little Léon into private *lycées,* with classmates like the privileged André Gide and with prizes in Greek and Latin. Maybe Blum was a born Socialist. At the tender age of five, he asked his capitalist father how he could bring himself to sell ribbons for more than he paid for them. As a matter of record, Blum was converted to Socialism in college, by Lucien Herr, librarian of the ultra-conservative Ecole Normale Supérieure (radicalism must have been in the air everywhere), who also converted Jean Jaurès and the poet Charles Péguy. After graduating from the Sorbonne, where he majored in literature and law, Blum, then twenty-three, precociously became a member of the Conseil d'Etat, whose judicial position is among the highest in the country. It is devoted to upholding, among other legalities, the French system of private property and capitalism, which Blum dialectically wished were in limbo. He continued this career in the magistracy for twenty-six years. His only known radical rebellion was the wearing of a derby to court instead of the regulation top hat.

As a member of the magistracy, Blum at first received an annual salary equivalent to eight hundred dollars, so, to earn a little more money, he became a journalist in his hours outside the court—specifically, a horse-race columnist and a book and theatre critic. He began to frequent the finical literary salons of Countess Anna de Noailles, the poetess, and of Mme. Caillavet, Anatole France's mistress. While still in his early twenties, he joined the staff of the *Revue Blanche,* a staff that included Marcel Proust and Gide. It became the great intelligentsia monthly magazine of Europe, with contributors like Verlaine, Mallarmé, de Gourmont, and Anatole France. Debussy was its music critic. Together, Blum, Gide, and Proust fought for Ibsen, introduced Tolstoy to the French, and claimed to be intellectual anarchists. "Intellectual anarchism" was a fashionable parlor-pink term, which later they sheepishly said meant, mostly, anti-Victorianism. Just as Blum's years in the Conseil d'Etat gave him a knowledge of the technical functioning of the French state, so his early bookishness gave him a knowledge of the shifting social climate of the French mind; both were invaluable to him in party politics and in his work as a political reformer. He thus dually developed the prodigious braininess that was characteristic of the Third Republic's outstanding political leaders.

Blum was the last of the notable politicos who in their youth participated in the Dreyfus Case, which in the late eighteen-nineties convulsed France, bewildering the Parliament and the people, making a fool, knave, and liar of the Army, overthrowing ministries, and almost toppling the Republic. It started with the Army's erroneously and illegally convicting Captain Alfred Dreyfus—the first Jewish officer ever appointed to its General Staff—of selling military secrets to the Germans, and sentencing him for life to Devil's Island. It developed into one of the most tragic, funniest, and most violent political medleys of spies with false mustachios, gullibilities, idiot experts, loyally lying officers, and dangerous confusions that any modern civilized European state ever suffered through, with the Army, which had started by convicting Dreyfus, finishing by being itself convicted. (It also led, later, to the disestablishment of the French Catholic Church.) Blum served as an unpaid, volunteer legal clerk to the lawyer who defended Zola in the libel suit brought against him in 1898 for his famous "J'Accuse," the Dreyfusard pamphlet that finally split the Case, the army, and France wide open. With the wretched Dreyfus, still in chains on his Devil's Island, as a focal point, two diverging lines cut deeper and deeper into French society. The anti-Dreyfusards, anti-Semites, anti-republicans, conservatives, diehards, Army, and Church were massed on one line, claiming the existence of a wicked Jewish syndicate. On the other line were pro-Dreyfusards, republicans, liberals, libertarians, anticlericals, anti-militarists, and believers in *les droits de l'homme*. The Right and Left division then drawn still cleaves France. The fairly young Army officer Philippe Pétain, on duty in Paris, was on the anti-republican side, which eventually led him, by then Maréchal, to Vichy, where he ordered the Riom trial of Léon Blum. In "Souvenirs sur l'Affaire," his book on the Dreyfus Case, written years after it was over, Blum wondered, in one of his best-known paradoxes, if the finally vindicated Jewish captain, who, ironically, turned out to be a typically dull, confused, loyal, military-minded man of his time, would have been a Dreyfusard if he hadn't been Dreyfus. The Dreyfus family never forgave Blum.

Blum's books usually plunged him into trouble. Published when he was twenty-nine, his "Les Nouvelles Conversations de Goethe avec Eckermann," a fantasy in which the Sage of Weimar plans a new character for Faust as a modern Socialist French deputy, was stimulating enough to annoy many French. "Du Mariage," which appeared when he was thirty-four, declared his opinion that permitting sexual experience for women before wedlock, as society permits it for men, would make for solider marriages. This incredibly advanced feminism naturally horrified the French, especially French mothers. Even the Socialists wished Blum would merely talk, and not publish. Blum was one of the early enthusiasts who helped reëstablish Stendhal at the beginning of the century. In 1914, his "Stendhal et le Beylisme" appeared. The trouble this book stirred up was purely literary.

In the summer of 1914, the assassination of Jaurès, Blum's friend and god and the dominant chief of the Socialist Party, belatedly pushed Blum, almost as though he were performing a penance, into Socialist Party work, which he had always refused to engage in for Jaurès. On the eve of the First World War, the powerful German Socialists hastily sent delegates to Paris to ask the French Socialists a question that millions of frightened European workers thought could elicit the miraculous answer on how to prevent a war: Should the French and German Socialists combine—dragging with them, perhaps, all the other workers—and pacifically refuse to be mobilized? When the talks were over and war had been declared, both the French and the German Socialists patriotically stepped into uniform, and the vaunted major ideal of seventy-five years of anti-military European Socialism was drowned in blood. Refused by the Army because of his bad eyes, Blum took the post of executive secretary to the Socialist Minister of Public Works. In 1919, the carnage ended, he was elected Socialist deputy from the Seine, and he finally began his Parliamentary career.

Almost immediately, Blum performed one of his two greatest feats, the significance of which he certainly could not have known at the time. Just before 1920's most important postwar Party congress, which was to be held in Tours, French Socialism, which still had a family relationship with Russian Marxism, began wondering what on earth had been going on in Moscow since the bloody Bolshevik Revolution—and how about Moscow's sending complete details? In answer, Moscow sent a sharp inquiry as to what on earth poky, unrevolutionary French Socialism had been doing, especially about establishing the dictatorship of the proletariat. On top of this, the Bolsheviks sent a dictatorial telegram—it was later known by the name of its signer, Grigoryi Zinoviev (afterward liquidated)—preparing the way for the laying down of twenty-one autocratic points the French Socialists, along with Socialists everywhere else, would have to concede to if they wanted to be admitted into the new, glorious Third International. One point was that they would have to throw out of their Party the French Socialist Charles Longuet, who was the husband of the unhappy Jenny Marx, and so Karl Marx's son-in-law. During the ensuing wild meeting in Tours, Blum led the opposition to the Bolshevik demands, and lost. Three-fourths of the delegates, led by Marcel Cachin, walked out, to found what is now the French Communist Party, and, as the majority, legally took with them the Socialist Party's newspaper *L'Humanité,* founded by Jaurès, and the Socialist Party's funds. Blum was the leader of the Socialist Party's penniless remains. He became, at that moment, its brains. He was then forty-eight. Without him, his historic rebellion against Cachin, and his minority, the leadership of the organized French working class might have become predominantly Communist after the First World War, instead of after the Second. Under Blum, France's working class became predominantly Socialist. His action gave the Third Republic twenty years of grace.

The electoral triumph of Blum's Socialist-dominated Front Populaire in the nineteen-thirties was, like all the reform periods in France, a reaction against a period of excessive political corruption. Financial scandals had shaken the country. In 1933, Center and Right governments had fallen like leaves from a calendar as their politicos' complicity was established with the notorious swindler Serge Stavisky, who had sold forty million francs' worth of worthless bonds to the French people. On the night of February 6, 1934, thousands of outraged Parisians of all classes rioted in the Place de la Concorde, shouting, *"A bas le Parlement! A bas les voleurs!"* Troops fired on the crowd, and seventeen or perhaps eighteen citizens were killed. Paris was in an uproar. Over the weekend, groups of Fascist *Croix de Feu* and the royalist bully-boys, *Camelots du Roi,* roamed the streets. Parliament and the very theory of Parliament descended to their lowest status in history. In reaction, finally, rose the notion of a Front Populaire, or bloc of Left Wing parties, which were to be united in reform—and ambition. As chief of the Socialist electoral campaign in 1936, Blum, at the age of sixty-four, became a familiar sight to thousands of French who had never laid eyes on him before. Especially in the turbulent, smoky mass meetings that packed the gigantic Vélo d'Hiver, Blum became a star attraction—a magnetic, though in some ways ill-equipped, political platform speaker, physically well proportioned, medium tall, agile, and looking only middle-aged; in attire somewhat foppish, with pince-nez, a small walrus mustache, and gray spats; and with a high, fastidious voice and the diction of a literary purist. These last two qualities were offered without embarrassment to listening factory hands accustomed to noisy demagogy. In the general elections, the Socialists won their first great victory. With a hundred and forty-six deputies out of the Chamber of Deputies' six hundred and eighteen, they became the largest party. In June, Blum was made Premier of his Front Populaire government, supported chiefly by Radical Socialists and Communists. The Front Populaire was Blum's second great achievement, even though it was, on the whole, a failure, being too experimental in form and too mixed politically.

For the workers, Blum's great social reforms were the forty-hour week and paid vacations, innovations that embittered the French owner class. In what he considered were the interests of peace, and perhaps at England's insistence, he refused to intervene in the war in Spain; this soured many European idealists, including members of his own party, and it infuriated the Communists. On taking office, he had meticulously stated that his was not a Socialist government but a mixed government wherein pure Socialism, in fairness to his non-Socialist partners, could not be put into practice. One year later, he was overthrown by the conservative Senate, which refused him the extensive fiscal powers that the Chamber of Deputies had just voted him. He and the Front Populaire, in various versions, dragged on into 1938, but it had long since shot its bolt. In 1936, surrounded on the Boulevard St.-Germain by a mob of *Camelots du Roi* who were accompanying the funeral cortege of the royalist historian Jacques Bainville, Blum, who chanced by, was dragged from his car and attacked, and his face and neck were badly cut. Inspired by the class hatred his regime and reforms aroused, the rich accused Socialist Blum of being rich, of having a fortune hidden in Swiss banks, of possessing a fabulous collection of silverware in his modest Ile St.-Louis apartment. This last rumor, he patiently explained, was the result of his having bought some second-hand packing cases to store books in during the First World War. The packing cases had previously been used for silverware and therefore were marked *"Fragile. Argenterie."* None of his enemies believed him. The old Dreyfus Case cry of anti-Semitism had been revived after he became Premier. Indeed, in the days preceding the election won by the Front Populaire, certain conservative and Fascist French had openly shouted, "Better Hitler than the Jew Blum!" Starting in June, 1940, Paris, and then all of France, had Hitler, and had him until August,

1944. Blum was arrested by the Nazis.

After being imprisoned in various châteaux, Blum, early in 1942, appeared before Pétain's special Riom court, created at Hitler's demand; he was accused of having helped cause the military defeat of France by inadequate rearmament during his Premiership. Blum's fearless, witty protests that he had at least provided all the armament requested by Marshal Pétain's inadequate armament program, and his attacks on the legality of the court, both backed up with similar statements and attitudes by his fellow-prisoner, ex-Premier Edouard Daladier, resulted, after twenty-four sessions, in the accuser Pétain's seeming to become the guilty party in the trial—or so Hitler commented in disgust as he called the trial off.

Following another period of imprisonment in France, Blum was sent to the German concentration camp of Buchenwald early in 1943, bravely accompanied by the lady who went there to become his third wife, the first two having died. Blum was so walled off in a private house on the edge of the camp that he did not know anything about the horrors around him till months later. In "Le Dernier Mois," his last book, he recounts the shocks and confusions of being carted around by the S.S. for three weeks in April, 1945, from one camp to another—Regensburg, Schönberg, Dachau, Innsbruck, Präger Wildsee—as his guards fled with their valued prisoner before the oncoming Americans. Blum was suffering from sciatica and from the pain of knowing that his life could be saved only in trade for immunity for Martin Bormann, chief of the Nazi chancellery. Blum and his wife were rescued in May, a few miles from Dobbiaca, Italy, near the Austrian border, by the American Ninth Army.

In the summer of 1949, two months after an intricate intestinal operation, Blum was again receiving political and press visitors at Jouy-en-Josas with his same old spirit. Attired in a fine green dressing gown and scarlet silk neckerchief—he was always an elegant, if eccentric, dresser—he repeatedly bounded from his couch, as if fresh ideas acted on him physically, even in his invalid state. His voice had grown thinner and more like a high reedy musical instrument, but his mentality was vigorous and flourishing. One afternoon, among other things, he said sadly to this writer that what he had wished to see in postwar France was, naturally, a labor government like England's. He reaffirmed his faith in traditional Socialism—no revolution but evolution and gradualism (the Socialists' basic law is gradual change), and free speech for all. He loudly, scathingly denounced the dictatorship and terrorism of Communism, as if he wanted his enemy the French Communist Party to hear him. He spoke of France more with reminiscence than with immediate hope. He felt that the Fourth Republic was merely repeating the Third Republic, as if French history had begun, in a senile way, to stutter its old thoughts, having refused to learn anything new—and appropriate to the middle of the twentieth century—after France's most terrible, most instructive war.

Surrounded by friends, visitors, and his books, Léon Blum, sprightly and hollow-eyed in his handsome dressing gown seemed an intimate historical figure, an erudite Frenchman who had seen history, made history, and suffered from history.

March 29

This has been a remarkably quiet, orderly two weeks of strikes, well handled by the government and the unions. Both have had plenty of experience. Strikes, which have been legal in France since the law of March 21, 1884, are called *grèves,* for the Place de Grève (now the Place de l'Hôtel de Ville). The Place de Grève was the old city center where workers came to get jobs or to protest, usually unsuccessfully then, against low wages. The recent strikes have been of the same type—strikes for more pay—and not political strikes, which doubtless would have included violence.

These quiet strikes brought quietude indeed to Paris, beginning as they did in the transport services here before spreading over the whole of France and into most of the major industries and bureaucracies. At the Gare St.-Lazare, normally the busiest station for commuters, the uninhabited glass-arched open-air train shed and the empty converging tracks looked as lifeless as a draftsman's sketch. Underground, Métro trains rumbled around town occasionally, and those who rode on them rode free, because ticket sellers and ticket takers had been called out. As substitutes for the city's green buses, which were totally immobilized, Minister of Transport Pinay supplied a few Army camions, operated by white-gaitered soldiers, to pick up passengers on the Concorde, and installed here and there a spasmodic service of those luxury autos usually seen filled with white ribbons, flowers, and wedding parties. (The ex-wedding cars charged more than the buses, which vexed the French, who remain economical even in a national emergency.) The government also asked Parisians driving their cars to give lifts to the rest of us, hopefully standing on street corners in the March wind and rain. Mighty few gave anyone a lift, even down the Champs-Elysées, which was jammed morning and evening by erstwhile Sunday drivers driving to their offices. As the number of Paris taxis is limited by the high price of gasoline and is inadequate even in normal rush hours, most of us walked during the strikes. On the whole, young women managed the best. They fell back on their German Occupation experience and bicycles, and independently scorched all over town, or they dug out their ski trousers and boots, which were all the style here that cold winter after the Liberation, and strode off to work looking as if they were headed for the mountains.

The strikes, of course, killed the Easter-holiday plans and non-motoring foreign tourist trade. On Good Friday, a group of lucky English tourists stuck in Calais by the railway strike were invited by a young French locomotive engineer, who was headed for Paris to see his wife, to climb aboard his unscheduled train if they were going his way. Half way home, he ran out of coal, and he refuelled at some unattended, strikebound railroad coalyard. He eventually larruped into the Paris *gare,* with himself and the English pleased as Punch. The strikes were not unpopular, even among Paris non-strikers, who, with a devotion to work as disciplined as the strikers' devotion to non-work, rose by the hundreds of thousands an hour earlier in the morning and bedded down an hour later at night, so as to walk from one end of Paris to the other, be on the job on time, and complete their regular stint—almost uselessly, since almost no business was transacted. When the government finally announced that, according to its statisticians, the cost of living had gone up twelve and a half per cent since August, the non-strikers also heaved sighs of relief, for it seemed as if their plight, too, had been noted.

April 26

Although Parisians are familiar, by hearsay, with the untrammelled, picniclike public spectacle called a New York welcome, they were downright bewildered by reports of the millions of waiting people, the acute emotionalism, and the extreme partisanship that were assembled on Manhattan's streets while ticker tape streamed from skyscrapers, in the greatest and most frenzied ovation in the city's history—all for General MacArthur. About the only thing the French are sure of is that it could not have happened here. With such tension and such differing views and with even a fraction of such a crowd on the Champs-Elysées, paving stones, not paper streamers, would have been thrown, and the result would have been a bloody riot. Tempers and sensibilities seem to be keyed up everywhere these days. The Paris papers say that Marshal Pétain (at the age of ninety-five still France's most controversial military figure) is probably dying in his fortress prison on the distant Ile d'Yeu. At the mere mention of his name this week in the Chamber, such a storm of outcry, invective, and desk-banging broke out that what the deputy who had the floor was trying to say was never heard, and neither was the voice of Speaker Herriot trying to restore calm.

The Truman-MacArthur crisis and the vast American civilian population's determination to decide for itself which man's military strategy will best serve history has certainly caused worry here, but it has caused no confusion of opinion as to which man is right. The French are solidly for Truman. Not one Paris newspaper has taken anything but the President's side. All the papers—except, naturally, those put out by the Communists—eulogized the General for his earlier military exploits, admired the careful omission of incitement in his address to Congress, which they think was simply a more magnificently worded version of his previous garrulous insubordinations, and disagreed, in real alarm, with his thesis on expanding the war in China. What has become in the United States the Greater Debate—between Republicans and Democrats arguing about war in the Far East—to the French still seems a matter that concerns not just two inflamed American political parties but at least ten of the sixty United Nations and the chances of life or death in Western Europe.

On July 8th, Paris will be two thousand years old and will begin her celebration of the *Bimillénaire,* which this year is the town's biggest summer fête. Julius Caesar conquered Gaul in 51 B.C.—actually two thousand and two years ago, of course. July 8th has been haphazardly chosen as the birthday because the weather ought to be nice.

June 20

In the elections just held, the first French national elections in five years, the citizens have voted in six and a half different directions, everybody hoping to get, from one direction or another, some things he wants very badly. Studying what it is the people want and how many of them, according to the urns, want it, is perhaps the easiest method of reading the election results. There were six winning parties, or groups, figured on the basis of seats gained in the National Assembly, and of these General Charles de Gaulle's Rassemblement du Peuple Français is on top, with a hundred and seventeen deputies sitting for him among the Assembly's six hundred and twenty-seven. De Gaulle's victory, much feared and much hoped for, has been achieved, but it is no landslide, and, since it may turn out that the government will be formed by a coalition of Center parties, his extreme Right-wing party may not even participate. Moreover, because he owns no daily French newspaper to come out with headlines saying he won, no paper has said it; in fact, all the papers noted his winning total obscurely—the conservative *Figaro* putting him at the bottom of the list. The things badly wanted by the fifth of the French electorate that voted for de Gaulle are order, a chance to put France back on her feet (and on a pedestal), a strong hand against pressure from the United States, the suppression of the Communists in the French government (his party, in its special language, calls them "separatists"), and, above all, that *mystique* of single, male leadership that European history has so often recorded and that this century has had considerable sad experience with. Indeed, what de Gaulle's voters want from him is precisely what he has, so far, offered—history and himself. In an electoral radio speech, he spoke of History, with a capital "H," having imposed on him "the responsibility of intervening to show the path and lead the nation. . . . After the difficult victory of 1945, I gave the republic back to the people. . . . The R.P.F. is formed around me for the good of the people." Something his working-class followers also want is his Capital-Labor Association program, despite the fact that he has never told them much about it, and despite the fact that at the first hint of it capital knew enough to dissociate itself from him in horror. An R.P.F. booklet presents the Capital-Labor

Association (it opens by mentioning first the French phalansterian Fourier and then Pope Leo XIII's famous encyclical on social justice, "Rerum novarum") as being against the class struggle and in favor of profit sharing by employees, as a "new" wrinkle in employer-employee relations, but with strict managerial authority and with protection of capital investment. His voters also appear to want, as part of their *mystique,* the picture they get, at a distance, of his very exceptional character, including his egotism (to them it seems a distinguished inheritance rather than a political acquisition), his personal probity, his majestic vocabulary (taken from the religious orator Bossuet), his heavy, dynastic-looking face, and his reiterated views on legitimate republicanism, according to which, if he had won his landslide, he would have aspired to be elected, through the legal devices of referendum and dissolution, President of France and to function also as Premier, a combining of offices that at the moment is forbidden.

The liveliest part of the campaign was the battle of the billposters—a political-advertising technique new to everyone but the Communists and really an imitation of the very successful posters that have been put up since early this spring by the anti-Communist organization calling itself Paix et Liberté. That organization's prize campaign poster was a vivid, Dufy-colored one of the Tour Eiffel, shown rocking under the weight of a hammer-and-sickle flag flying atop, and bearing the warning *"Pas ça!"* The Communists considered it such effective anti-Communist propaganda that they hastily printed American flags to paste over the Soviet emblem. De Gaulle's tricolor posters were the noblest-looking and the most artistic typographically, his publicity man being the novelist André Malraux, recently turned art authority. One showed the tortured, handsome face of La Marseillaise (from one of the François Rude statues of the Arc de Triomphe), with the words *"De Gaulle vous appelle, pour que la France soit la France."*

The elections were held on Sunday, June 17th. Exactly eleven years before—on Monday, June 17, 1940—Maréchal Pétain announced to the French over the radio that France had fallen. Paris newspapers did not mention the significance of the election date, and most of the voters seemed to have forgotten it. They remembered the Maréchal himself on Sunday night, though, when the radio announced that the very ill prisoner was to be transferred from the Ile d'Yeu fortress to a mainland *résidence surveillée,* where he would be hospitalized. On the anniversary of what he tragically began then—the dilapidation of French republicanism—nearly half the electorate voted for the two political extremes of authoritarianism.

The Paris bimillenary celebration, or two-thousandth birthday party, has so far been a struggle between hedonistic outdoor fêtes, arranged against historical architectural backgrounds, and the grim weatherman. The magnificently conceived Louvre Court concert was called on account of rain, like a baseball game. (It was given ten days later.) The Henri IV concert, held in his Place des Vosges, with its *souper* under the arcades, for living French nobles personifying their own or somebody else's ancestors, was interrupted by the worst thunderstorm in even the common people's long memories. The historical art shows, which are being held indoors, have proved safer. The expositions have been extraordinary in their variety, their specialties and oddities interesting both residents and tourists, and the organizing imaginations behind them deserve a vote of thanks. The following is a sketchy listing of what one can see if one's eyes and feet hold out: Charpentier's rich Plaisir de France paintings; the Palais de Glace's First Salon of Hunting and Venery, which is fascinating; pre-Gobelins tapestries, shown at the Musée des Gobelins; the art of glass, at the Pavillon de Marsan, which is fine; Conquests of Photography, at the Muséum d'Histoire Naturelle; the Palais de New-York's show of women painters and sculptors, which is only pretty good; chefs-d'œuvre of the Louvre's French school, at the Petit Palais; Napoleon and his Family, in a show of arms, trophies, uniforms, souvenirs, and so on, at the Invalides; Sèvres table services, in Sèvres; a great Toulouse-Lautrec show, including unfamiliar early, serious works, at the Orangerie; and, if one can get to the suburban Sceaux between showers, an exposition of water colors of "the environs of Paris," beginning with Corot and coming down to the present. Horse racing and steeplechasing have started, hotels are jammed (principally with North and South Americans, most of whom have brand-new Paris umbrellas), the great Paris buildings and squares and fountains are floodlit every night, prices are high, and it is a big tourist season.

Jean-Paul Sartre's new play, "Le Diable et le Bon Dieu," at the Théâtre Antoine, is the season's "must" play. It takes four hours to perform. Sandwiches are sold between acts, as if the Antoine were

Bayreuth. Sartre's theme is the antique one of man's terrible indecision about whether to be good or evil, combined with the more modern one of where good and evil themselves come from, and he did well to put the Devil first in his title. For Goetz, his rich-fleshed, rich-minded German warrior protagonist, fighting with men in battles near the city of Worms and with his conscience on the edge of the Reformation, is more alive as a man, more interesting as a character, wittier to listen to, and far better as a stage figure when the Devil is leading him, in the first act, than he is, after he has turned to God—literally on the throw of dice, in the spirit of a gambler changing his system—in the subsequent two acts, which are lifeless, loquacious, and argumentative, and include eleven tableaux. Goetz is really a dialogue in himself—Shavian, shocking, blasphemous, coarse, and human in Act I, and then, in the following acts, talkative in the dreary medieval, Maeterlinckian theatrical manner as he hunts for the good. Goetz is never Faustian, at least; there is no mean merchandising of an old soul for youth and pleasure. Goetz is apparently meant to be a premature modern, with an outstanding talent for doing and enjoying evil, who, as the result of a cerebral whim, changes his métier to following godliness, for which he lacks the genius that is faith. He returns, finally, to the Devil and butchery, with the excuse that goodness does not pay on earth and that God has not let him work a miracle—like a trick of legerdemain—and this top-ranking blasphemy seems the only conclusion that could bring the curtain down. It is admirable that the philosopher Sartre has been able to put into a theatre the Manichaean problem of good and evil in man, but certainly something is the matter with the play as a play. It arouses outspoken dissatisfaction in everybody, and everybody goes to see it. Pierre Brasseur, as Goetz, gives one of the greatest Act I performances of our time.

July 19

One of the greatest summer art shows of the many given in recent decades by the French government is "Le Fauvisme," currently at the Musée National d'Art Moderne. The Fauves, or Wild Beasts (their brilliant colors supposedly snarled), received their name in the Salon d'Automne, in 1905, when the paintings by Derain, Vlaminck, Marquet, Van Dongen, and so on looked much the way the paintings of their leader, Matisse, still look—some with broken green brush strokes like fallen leaves, and all with the savage shock of the bright colors that attract civilized man's eye to nature's most exciting flora and fauna. There was also visible a determination to wipe out the studio painting of the past and start fresh, for the nth time. Before the Fauves, the Impressionists were damned, and after them the Cubists. France has specialized in violent political revolutions; these groups were responsible for revolutions in modern art, with vermilion blood shed on the palette.

The Fauve exhibition has an impressive scope, which includes a preliminary explanatory room of paintings by the influential pre-Fauve artists Manet, Gauguin, van Gogh, and, above all, Cézanne, who was little thanked at the time. Then the show moves into its strength in the wonderful first-rateness of the hundred or more Fauve canvases—Derains of his vivid Collioure period, Braques still as gay as zinnias, Seine scenes of running purple water by Vlaminck, lovely Camoins, Marquets, and Manguins, and fifteen of Matisse's greatest, most famous paintings. Intense interest has been shown in his Pointilliste 1903 "Luxe, Calme, et Volupté," which once threw Parisian art-lovers into a rage; the 1906 study for "Le Luxe," with its strange duck-egg-colored female nudes; "La Gitane," with her green breasts and nose; a 1907 portrait of Mme. Matisse that looks like a Picasso of 1922; and an 1899 landscape of the city of Toulouse, featuring the innocent pink-lavender that the Wild Beasts all later used— the naïve rainbow color that European man had shut his eyes to as being in a kind of suburban bad taste.

There is still enormous pleasure in viewing these great beginnings of twentieth-century modern art in Paris, where the colors became part of the gray city's brighter history—these pictures produced before the Wild Beasts moved into duller colors and stiffer formulas, leaving old Matisse today the last, as he was the first, practicing Fauve.

August 1

It is more than a week now since ex-Marshal Philippe Pétain died, an unpardoned nonagenarian state prisoner on the Ile d'Yeu. Some few people thought his death could be tinder for trouble in France, though just what dangerous phoenix might rise from his ashes they seemed not to know. The announcement of the death itself, which had been long expected and which he long resisted, was received by the vast majority of Parisians with flat calm, some indifference, and a feeling of general relief, both for him and for themselves. He was so very old. As long as he still breathed, whenever they read or even thought about him, which in the past few years was admittedly not often, he symbolized the corrosive, demoralizing years of occupation that France's citizens wanted to forget and that most of them, perhaps to an alarming degree, have succeeded in putting out of mind. By his death on an island few ever now, he was finally disposed of at a distance, separated at last from their postwar existence in France, where their political emotions, interpretations, troubles of conscience, loyalties, disillusions, and memories of him good and bad reached their logical climax of partisanship and grief in his trial for treason, a courtroom condensation of all humiliations for them, for France, and even for him, the Verdun general, sobriqueted victorious, of a former, better war. No nation could rise to the same pitch twice without additional, reanimating events. Pétain's dying brought little of it back, except in the newspapers.

Had Pétain's funeral rashly been held in Paris, the French, who are sentimentally and politically sensitive to such black-trimmed processions, might have reacted more positively, at least for those few brief hours. Even his well-known wish to be buried "at the head of my soldiers" lying at Verdun—a solemn desire that to many French who no longer bothered to be bitter about him seemed a permissible last indulgence, and that to some seemed almost a command from the deep past where they had first heard of him, with gratitude then—failed to fulfill newspaper prophecies that "Pétain burial honors will split France." Except for his former trial lawyers, still briskly and unsuccessfully pleading on his behalf any cause they can find, and the small remaining emotional band of demonstrative pro-Pétainists, the public accepted without demur the official statement that no one had the authority to allow his return to the mainland of France, living or dead. The ex-Marshal's death and burial on an island gave him, in very different historical circumstances, a greater exile, and far less legend, than St. Helena furnished to Napoleon.

One thing about Pétain's death is sure. Of the many dicta he enunciated in his trembling old radio voice, so often heard from Vichy, the one the newspapers quoted most often, as substitute for *de mortuis*—and the one the French most resent today—was his statement that history alone would judge him. Probably seven or eight out of every ten French were automatically pro-Pétainist at first. It is a similar huge aggregate, now anti-Pétainist, who think that history already has judged him, and that they, the French people, were the historians. They think that they have lived his history, for they heard a lot of it from his trial, despite his silence, learned even more about his Nazi colleagues from Nuremberg and Dachau, and can hear further volumes just by listening to their own memories. They think he chose wrongly at every turn, and that a leader's choices, not his intentions, are what write the line of a nation's history. Nearly without exception, the Paris papers listed the main tragic chapter headings—1940, *le shakehand* at Montoire with Hitler and "collaboration with honor," which included anti-Semitism; 1942, the command to resist as aggressors the American liberation forces in North Africa, and concurrent permission for Nazi troops to take over his Free Zone on their way to seize the French fleet at Toulon; 1943, when the Grand Maquis was forming, his radio voice quivering, *"Parents français!* Do not give your children bad advice [i.e., do not encourage them to join the anti-Nazi resistance]. Counsel them to go work in Germany." Apropos Verdun, even his press obits here mentioned what it has taken two wars to make generally known—that if he bravely bled France at Verdun, he also advised retreat and was withdrawn by General Joffre, leaving General Nivelle to say what were supposedly Pétain's most famous words, "They shall not pass," for it was Nivelle who was in charge to win the final hemorrhage of victory.

For those of us who saw Pétain at his trial, the visual memory of him is of his sitting silent in the belated dignity of total uncollaboration at last—with the Palais de Justice assemblage, in any case. From time to time, he lifted his bare hand and flapped his gloves like a fan to protect his face from the myriad focussing retinas of cameras, as if they were multiple-eyed modern insects. An anti-republican, he disapproved of any republic. It is characteristic of the recent republics here that there was no government in office for the important event that was his death (nor any government since, either), but also typical of the viable republicanism of the French people that they have easily met both happenings without losing stride.

One of the important events in the lively posthumous literary career of Henri Beyle, who changed his name to Henry because he liked the letter "y" and to Stendhal because he worked for a French government that did not like political-minded authors, was the Academy's award of this year's Grand Prix Littéraire to Henri Martineau, France's foremost Stendhalian. It was given to him for his thirty years' labor of deciphering and annotating Stendhal's madly illegible manuscripts; for the new minor volumes and text expansions that resulted; for his re-editing and correcting of all earlier printed works, the first complete, definitive Stendhal, published in seventy-nine volumes; and for Martineau's voluminous explanatory writings over the years, of which the latest is his new tome, "L'Œuvre de Stendhal: Histoire de Ses Livres et de Sa Pensée." Martineau, whose blue painted Librairie du Divan is a landmark on the Place St.-Germain-des-Prés, was a Vendéen doctor until he took up with the Bibliothèque de Grenoble, the city where Beyle was born, which he hated and where his manuscripts have come home. Only fourteen of Beyle's books were published in his lifetime. Grenoble has thirty-two signed manuscripts, plus masses of loose papers and marginalia (such as his annotated copy of Destutt de Tracy's "Commentaire sur l'Esprit des Lois de Montesquieu," recently bought in America) in which he indulged his passion—almost as great as his love for secrecy—for scrawling enigmatic notes about anything on whatever was handy. His opening plan for "Vie de Henry Brulard," another of his pseudonyms, since it is really his autobiography (in his notes, though, he always referred to himself as Dominique), is scribbled in his copy of "Clarissa Harlowe." Martineau says of all the material in Grenoble now there is nothing left unpublished but a handful of leftover pages of "Lamiel," Beyle's unfinished last novel, whose final five hundred words, rushing from his pen a few days before he collapsed in the Boulevard des Capucines, were written with such speed, sickness, and malformation that only their general intention has been deciphered.

For a novelist who became the first master of modern psychology at the wrong time; who was a known worshipper of Bonaparte but published his marks of genius, "Le Rouge et le Noir" and "La Chartreuse de Parme," under Louis-Philippe, no Bonapartish regime; who thought passionate writing "horrible style" and read daily at breakfast from the Code Napoléon, *"pour prendre le ton,"* and yet in those novels created two of the century's most passionate love affairs—for this kind of writer, who also rather ungratefully calculated that he might be appreciated around 1880, or perhaps even 1935 (writing in 1835 as we might write bitterly today of fame in

2051), Stendhal really had a lot of kudos and cash in his lifetime. These details Martineau sets down in full, along with other engrossing, less well-known, material. After all, "Le Rouge et le Noir" had favorable, if not enthusiastic, reviews. Stendhal received twenty-five hundred silver francs, or five hundred dollars, for writing "La Chartreuse de Parme" in seven glorious weeks. Its first edition sold out at the equivalent of three dollars a copy in a period when laborers were laboring for a few sous a day. Balzac gave it a seventy-page puff, still a record, saying, "Where I work in frescoes, you create Italian statues," but adding that the style was often negligent. And Goethe wrote in his "Conversations with Eckermann," "I do not like reading M. de Stendhal, but I cannot help it. I recommend you to buy all his books." Today's Book-of-the-Month Club's publicists might well wonder what more Beyle, with Balzac and Goethe as boosters, could have expected from the year 1880. In 1882, he was posthumously accorded Paul Bourget's famous essay that called him the modern master, and, sure enough, the fastidious cult of Beylisme took root. And in 1935 Martineau was ready to publish ten volumes of Stendhal's journals.

Martineau says Stendhal was pleased with his atrocious handwriting, which he hoped would keep people from prying into his secret papers. For further mystification, he also misspelled, using "k," which was another of his favorite letters, for "q" and "c," "s" for "z," "i" for "y," and also vice versa. He used mixed anagrams, muddles, and abbreviations, such as "Mero" for "Rome," "Téjé," with the stress on the final "e," for "Jesuit," "Tolikeskato" for "Catholiques," and the fraction $\frac{1}{3}$ as a pun for the politician Thiers. And though he never kept dates straight or added very well, he computed his age at fifty-three as $5 \times 10 + \sqrt{9}$. Martineau's book has a diverting chapter on "Lucien Leuwen," Stendhal's earlier unfinished novel, lately published for the first time in English in the States. Martineau told your correspondent that the "Leuwen" manuscript is "very instructive," rich in marginalia of gossip, pen portraits, and Beyle's advice to himself and remarks on his health, with asides, in passages on Leuwen's love life, concerned with how Italian coffee troubled Beyle's intestines, how champagne at lunch made him drowsy, and what he thought about wearing spectacles for the first time. As Stendhalians devoted to Beyle's peculiarities are fond of relating, he genially stole the novel's idea from a manuscript called "Le Lieutenant," featuring a Lucien Leuwen, that had been sent him for friendly criticism by Mme. Jules Gaulthier, one of the many ladies he loved and who understandably failed to fall in love with him. Beyle worked over her idea under various titles, including "Le Télégraphe," which had just been invented and which fascinated him. It is amazing how interest in his unsuccessful sex life and squabbles over who his character models were still agitate Stendhal scholars here. The two latest issues of the *Revue des Deux Mondes* contain articles on some vain vacation flirtations of his. The magazine *Le Divan* berates some Stendhalians' notion that Korasoff, Julien Sorel's Russian prince in "Le Rouge et le Noir," was founded on the Marquis Astolphe de Custine, a Restoration acquaintance of Beyle's and an aristocrat lately identified as an astute traveller; his anti-Russian observations were recently published as a book in America, with an introduction by Lieutenant General Walter Bedell Smith, head of the Central Intelligence Agency. Since 1882, the Stendhalian cult here has had time to grow pretty finical.

August 23

After thirty-two days of being without a government, the French find it nice to have one again. They think it ought to last just about thirty-two days, too. Actually, it will probably survive the current *petite* session of Parliamentary powwows and Parliament's seven weeks of autumn holiday. After that, say the de Gaulliste deputies, who already have named it "the vacation government," their General will push out Premier René Pleven's present setup and install his own. There has never before been such a huge Cabinet—a result of the government's effort to obtain a majority—as there is now, even in Léon Blum's Front Populaire regime, which had an Under-Secretary of State for Leisure. According to that cynical old cartoonist Sennep, this one includes several Ministries of Blah-Blah-Blah, or plain gab. It is a tragedy for French democracy, already split into pro-Communist and anti-Communist camps, that it now has a second violent split, on state aid to Catholic schools, with the anticlericals and Socialists bitterly against and the clerical Popular Republicans obstinately for—and with all parties glumly admitting that what France unquestionably needs is a thousand new schools of one sort or another. Today's *problème scolaire* is a revival of the question of the separation of Church and State, considered buried in the early nineteen-hundreds; is another stale battle in the Wars of Religion; is one more struggle between the two types of Frenchmen—a duality that has made modern France.

The death of Louis Jouvet was, if one may say so, the popular event of the month. It interested and affected the population of Paris to an extent that nobody—probably not even he, expert as he was on public reactions—would have thought likely. It was suitably dramatic. Stricken at rehearsal in his Théâtre de l'Athénée, he lay on a cot in his managerial office there for two days and two nights, in a speechless, recumbent pose, while completing his act of dying. Then the cot in the office became his bier, while thousands of Parisians, after work hours, stopped the traffic in the Rue Caumartin as they queued up to climb the little staircase backstage to give him, in their fashion, their final respectful applause. He was the only present-day intellectual French actor who was popular with the masses. Not once but day after day the newspapers devoted whole pages to him, retracing his career and attainments as if he had been a leading governing figure, which in a way he was—the governor of the French theatre, the actor-manager who acted and managed so that the great French classics, the fantasies of Giraudoux, the décors of Bérard became part of people's lives when the curtain rose. Giraudoux's most imaginative play, "The Madwoman of Chaillot," on which the author practically worked himself to death, was produced posthumously. Then Bérard was fatally stricken in the Théâtre Marigny. Now the third of the trio that made an epoch in the French theatre has died where he belonged—in the Athénée.

Seven years ago, on August 25th, Paris was liberated. The liberation itself began earlier on the streets, in the emergence of French men and women of the Underground. "L'INSURRECTION FAIT TRIOMPHER LA REPUBLIQUE A PARIS, LES TROUPES ALLIEES SONT A SIX KILOMETRES DE LA CAPITALE," the clandestine Resistance newspaper *Combat* jubilated in headlines on its first public copy to come from the press, on August 21st. *Combat* has just printed a photograph of that euphoric front page, along with some less happy up-to-date comments. *Combat* is, it should be noted, anti-Communist, even anti-Socialist, and Albert Camus, author of "The Plague," was among its first brilliant editors. "Since 1944," *Combat* commented sadly, "the world has changed. Only the inquietude remains the same. What seemed easy in August, 1944—liberty, security, social justice—has run up against the permanent obstacles of oppressive capitalism, intolerance, and privilege. We still have faith in man, in man only, in his good sense, in his good will." There are many, many French, even of Rightist political views, who, as they look back over those seven years, agree about these disappointments, which they think have come not from others but from their own leaders.

September 5

Bread is going up to fifty francs a kilo, a shock for the French, who eat about a metre of it a day.

Another bread shock has been even worse—the mysterious madness, agonies, and deaths from poisoned bread in the village of Pont-St.-Esprit, near Avignon, composing a medieval nightmare tale. It seems that the lethal bread came from Briand's Bakery, the best in town, on August 17th. The mystery was not solved until August 31st. By that time, four Spiripontains, as the villagers are called, had died in anguish, thirty-one had gone raving mad, five were in grave danger, and two hundred were sick. Even cats, dogs, and ducks that ate crumbs of the bread had fits. People seized with the affliction acted possessed; they saw monstrous visions, and burned with inner fires. Charles Graugeon, aged eleven, tried to strangle his mother. Mme. Paul Rieu needlessly attempted suicide, for she died anyway. Mme. Marthe Toulouse tried to leap into the Rhone to quench the fiery serpents inside her. One citizen fired his shotgun at the monster he thought pursued him. A peasant named Mizon, who missed buying his favorite evening *petits pains* at Briand's on August 17th, came by for them the next morning, went mad, and died on August 20th. In the insane asylums of neighboring Nîmes, Avignon, Montpellier, and Marseille, where the frenzied Spiripontains were sent, they frightened even the other lunatics. Only one photograph of a victim appeared in the press. It was of a M. Guignon, being carried to an ambulance, his legs and arms raised to fight off whatever he thought he saw, his elderly face, with open, distorted mouth, looking like a Gothic carving of an old man screaming in Hell.

The slow solving of the mystery, followed with palpitating interest by the French, was like a detective story. It started with a rumor of a mass political poisoning, developed into a fascinating study of the functioning of provincial authorities, and ended in a sort of Balzac formula. Professor Jean Olivier, chief toxicologist of Marseille, suspected poisoning from the chickling vetch, *Lathyrus sativus,* which produces lathyrism, or spastic paralysis. Then he discovered the presence in the bread of rye ergot, *Claviceps purpurea,* that little black abortion on grain heads that comes in wet weather and, it has been claimed, contains, among dozens of other chemical elements, about twenty alkaloid poisons, three of them virulent. The Montpellier 14th Brigade Mobile of the country police did a brilliant, slow job of deduction. Little grain is raised or ground in the Spiripontains' Département du Gard. There were twenty-three suspects, the mills in two nearby wheat-raising *départements.* Finally, one miller, Maurice Maillet, of the mill at St.-Martin-la-Rivière, in the Département de la Vienne, confessed that he had ground into his wheat granary scrapings of ergot-bearing rye, furnished to him by a dishonest local baker (it seems the baker's cashbooks were not even in order, which also shocked the French), who had bought the diseased grain illegally from unscrupulous peasants. Maillet, who now says his conscience pained him, sent the sack of bad flour to Pont-St.-Esprit "because I did not know anybody personally there." The Balzacian dénouement was, of course, cupidity. The two men were avoiding twenty francs sales tax per kilo on the grain, were dealing in spoiled goods for profit, and were using more rye in wheat flour than the law allows. Maillet has been arrested for homicide and "involuntary wounding" of others.

September 19

This is one of the moments when its being a small world after all is just what is making everyone so nervous. It was the shrinking of Europe's strength in two world wars and, in the last, the shrinkage of its size on the map that have necessitated the new intimacies on the Continent, the rallying of the Anglo-Saxon neighbors across Channel and Atlantic, and today's conferences in Washington and Ottawa, precisely planning (if possible) against a third world war that would shrink Europe to zero. It is exactly the smallness of this world, Parisians feel, and the terror of being reduced to nothing that have at last forced a kind of idealism onto the *conférenciers,* the ideal being a united Europe.

In Paris, there is a warranted feeling of pride and relief that at the Washington conference both the Americans and the British finally accepted the Schuman and Pleven European plans. They are French conceptions, they can be used as starters, in the lower materialistic depths of coal, steel, and permitted German soldiery, toward a basic communal Continental structure. They are also the only new international European notions anybody has thought up since the British put forth their balance-of-power scheme at the Congress of Vienna after Napoleon was defeated. As near as the French can make out (and there was one moment when the Washington Conference reached much a state that Premier Pleven had to tell Parliament he had just buzzed Schuman on the transatlantic phone to ask him to explain what on earth was going on), the acceptance of the Schuman-Pleven plans was due largely to the man who was not there. This was General Dwight Eisenhower, who frankly, almost optimistically, stated in July that he thought joining Western Europe together was the key to the whole thing.

With bitter realism, the French recognize—indeed, it is the Schuman-Pleven thesis—that the one proof of any federation will lie in the relations of the Big Three with the Bonn government. In other words, the developing destiny of what is left of Allied Western Europe will principally depend on its relations with the half of Europe's old attacker, Germany, that is left. This in particular makes it seem a remarkably small world. In a profound editorial, *Le Monde* just stated, "Politics and economics count less than the spirit. Europe will be no more if she does not forge a common soul."

A small book of a hundred and twenty pages entitled "Et Nunc Manet in Te, Suivi de Journal Intime" constitutes for French readers the final shock of the sort that André Gide's obstinate candor about his private life unceasingly furnished during his writing career. The booklet also probably constituted a shock to Gide's regular French publisher, La Maison Gallimard, for the new work has just come off the presses of the Editions Ides et Calendes, of Neuchâtel, Switzerland. To judge by the 1947 copyright, it was at the moment of his receiving the Nobel Prize, and also of his first grave alarm about his health, that he sent the manuscript abroad, to be published only after his death. Actually, thirteen copies were at that time privately printed, and distributed by him among his friends, according to his secret communicative custom.

The book consists of two parts. The first, "And Now It Remains to Thee," a tag from Virgil, was written after the death of Gide's wife, Madeleine, whom he called Emmanuèle in his many journals. She was also his cousin. This part of the book explains what he calls the secret conjugal tragedy of his life and of their existence together. The second part, the "Journal Intime," fills in all the blank spaces indicated by asterisks in those portions of his published "Journals" that deal with her between 1916 and 1938, when she died, thus amplifying also what was so obviously cut short in "Si le Grain Ne Meurt," a book of his memoirs up to the time of his betrothal. These revelations provide the inner sight without which, he declares in his posthumous record, the diaries are blind. This new book will unquestionably spill a great deal of ink here, from bitterly dissentient pens of literary critics, moralists, pro- and anti-Gideans, Catholics, Protestants, and perhaps plain, ordinary well-read French people, who may easily think that this whole recent confessional production might better have been blotted out. Gide's printed candor about his erotic tastes has long since ceased to surprise or shock his accustomed readers. In his posthumous book, it is the living man standing behind the admired novelist that will shock and surprise them anew. "Intoxicated by the sublime," or so he wrote in it, "convinced that the best of me was what communed with her," he made a white marriage, on the extravagant Puritan notion that only males are naturally carnal, wives being sexless—a special, civilizing dispensation of nature, which allows gentlemen to maintain intact the ideals of pure womanhood. With his wife, Gide recounts, he wanted an ethereal love, "the more ethereal, the more worthy of her."

Actually, the new book contains elements of a peculiar dramatic plot—the wife's silence and the husband's opposite, egotistic assumption, which combine into a human experience so unlike any hitherto intelligently recorded that even Gide, as an artist, apparently recognized that his story could never be believed as fiction but only as fact. The tragic high point for both Gides came when she set fire to all the letters he had written her over thirty years, which she considered "her most precious possession on earth" and which he loudly mourned as "the best of me, which disappeared and which no longer counterbalances the worst." "Perhaps there never has been a more beautiful correspondence," he lamented. The undeniably tender spiritual connection between them, or what could have remained of it by that time, thereafter grew feebler, without that earlier ink to live on. Madeleine-Emmanuèle died without their ever having discussed or mentioned what their lives together had not been. Gide was crowned as a literary figure in more ways than one during his lifetime. This last book adds thorns to his laurels—thorns that, because he is no longer alive to wear them, will cause pain principally to his admirers.

October 2

The winter art season has opened with an exposition at the Galerie de l'Elysée of Maurice de Vlaminck's oils. Here are his hallucinated landscapes, dreamy in overrich color beneath the strange, sugary-white frosting that always fills his skies. For the first time, he has added to his regular repertory, exhibiting a pair of odd journalistic still-lifes. One is composed of a revolver, a bottle marked "Poison," and an out-of-print Paris newspaper that used to specialize in crime news. The other shows a tin of salmon, a copy of *L'Humanité*, a knife, and some bread. Nobody in charge at the gallery seemed to know what these pictures mean beyond what they show, which is a revolver, etc., and a tin of salmon, etc.

As a young man, Vlaminck was one of the boldest of the Fauves, whose inflammatory colors inspired him to write in a youthful art magazine, "Ah, to burn the school of the Beaux-Arts with my cobalts and vermilions!" When he first took up painting, he was a music teacher (like all the members of his needy suburban family), a professional bicycle racer on the Paris-Bordeaux run, and a novelist whose works were published in the Orchid Books Series and included "D'Un Lit à l'Autre" ("From One Bed to Another") and "Ames de Mannequins" ("Souls of Dummies"), the latter a fantastic account of some wretched Russian émigrés living in poverty on the rim of Paris. Unfortunately, Vlaminck's own poverty was similar while he was doing the best painting of his life, as a young Fauve. Because canvases cost money and his paintings were then bringing none in, he wiped off many of his superb pictures of the Seine with a handful of grass and started others on the same *toiles*. A few of his Fauve paintings still exist, most of them in modern museums. The loss of the others becomes even sadder when one thinks of them in comparison with the almost lithographic landscapes he has been painting ever since.

November 29

The weekly *Figaro Littéraire* has just enjoyed the honor of printing three long fragments of Marcel Proust's early writings, theretofore unpublished, and, indeed, unpublishable until recently, when his energetic niece and heir, Mme. Gérard Mante, finished piecing these and other bits together into nine hundred pages—precious old scraps, one might say, strangely used to make a brand-new garment for somebody deceased. This new opus, which is entitled "Jean Santeuil," the name given by Proust to himself as the observer in the chronicle, was written between 1898 and 1910 and is clearly a preliminary use of material he later transformed into the romantic and creative carved Gothicism that begins, under the shadow of the Cathedral of Chartres, with "Du Côté de chez Swann." As the opening of the unfinished, unorganized "Jean Santeuil," Proust wrote, "Can I call this book a novel? It is less perhaps and much more—the essence itself of my life. . . . This book was not written, it was reaped." The *Figaro* fragments constitute about a tenth of the work. The first two are like long short stories.

"En Bretagne," the first of the series, contains one forty-line sentence that, like a layer cake, predicts the complex and more refined *madeleine* teacake to come. The "Bretagne" fragment also features that familiar summer-hotel seascape to come, this time in an imaginary village called Kerengrimen, where Proust, his sensibilities already stretched out like a telescope toward the lives of the other guests, focusses on the fatal love story of a middle-aged writer called merely C., who is prophetically like the mature writer-to-be P. For, as Proust suffered from asthma and jealousy, C. suffers from phthisis, a psychosomatic accompaniment of jealousy in love, which is his actual malady and of which he cures himself, only to die of his physical illness.

The second story, "Un Amour de Jean Santeuil," is almost the real thing, is almost pure Proust. Here he is close to "A la Recherche du Temps Perdu"—adolescence in the Champs-Elysées gardens, where the tantalizing Gilberte is tentatively incarnated as Marie Kossichef, a maddening little Russian, where the servant Françoise is nowhere visible among the trees, but where a St.-Germain duchess *is* visible. Proust's mother's bourgeois goodness has not yet been transfigured by her death, but her refusal to come to his room to kiss him good night when he was a boy, and his loneliness for her, which leads him to the miracle of hearing her voice over the long-distance telephone—these are there, though the literary style they were finally preserved in is absent. Proust's early style was rather like Flaubert's. The flamboyant Parisian style he derived, oddly, from Ruskin's "Bible of Amiens," from which English writers got so much less, had not yet been chiselled, just as Illiers, where his uncle lived, had not yet come to be called Combray but is called Etreuilles. However, all the elements of the anguished, carefully drawn out Proustian life, are there: the tempests of unrequited, jealous love, the slow, humid sufferings from rendezvous never kept, and the false, bright garden weather of elegant worldly life shining all around. "Un Amour de Jean Santeuil" is fascinating in its own right, and to Proust-lovers it is like an orchid plant, niched in the hothouse, and in leaf but not yet in bloom.

The third piece, "L'Affaire Dreyfus," must have cost Mme. Mante a lot of trouble to stick together. Its only coherence is the court reporting by Proust, then a young, well-to-do, observant Jew and genius, suffering both for the cause of justice and for the cause of race, who humbly carried sandwiches and a canteen of coffee to the Palais de Justice, where he daily attended the trial. From what he saw, he sketched his first really elaborate, full-length portrait. It was of the anti-Dreyfusard figure General de Boisdeffre, with his stiff leg, his shabby civilian topcoat, and his cheeks "embroidered with red like certain mosses covering autumn walls." Proust gave far less to his picture of the martyred Colonel Picquart, whose truthfulness saved the Dreyfus cause, seeing him only as a gallant spahi, dismounted and without weapons, walking into the ambush of the law.

General de Gaulle's R.P.F. Party has been worrying all the French who did not vote for him last June by its silence—by the fact that although it has the greatest number of Assembly seats, it has done little with them. At a recent three-day R.P.F. caucus on future policy in Nancy, where public speeches were made, the General and his right-hand man, André Malraux, both spoke without soothing anybody. Malraux received an ovation for describing the General as "the single figure toward which one day all the humiliated French will turn—all the innumerable humiliated French of yesteryear and the awesome crowd of those humiliated yesterday." The General himself said, "Because the Americans give us arms, we do not have to be their subjects." He also spoke against the creation of a united European army before there should be a politically united Europe, and was hostile to the Schuman Plan for the European coal-and-steel pool. He made it clear that his R.P.F. deputies would fight both these pillars of French foreign policy when they come up soon in the Assembly. "It is probable that under the formidable pressure of events, common sense and patriotism will lead others onto the road that we have taken," the General said. "I declare that we are not inclined to refuse any helping hand. We aim at serving the country but certainly not at having a monopoly on good ideas or good actions." He also called for constitutional changes that would strengthen the executive branch of the government and his party's position in the Assembly. So now the French know the General means to do one of two things with his dominant party. In a new turn in policy, he means to support any political party that agrees with him and he means, as he has been understood to mean in the past, to take the leadership of the Assembly, if he is given enough power so that the other parties, hot or cold, would *have* to agree with him.

December 5

An astonishing concert of ultramodern music was recently given in the gloomy little Salle d'Orgue of the austere Conservatoire. There were songs set to "Mélodies en Langue Imaginaire," by Mlle. M. Scriabin, a relative of old A. Scriabin, whom she has left far behind. The chief interest, however, was in two examples of quarter-tone music—"Fragment Symphonique" and "Fugue No. 1," composed by Ivan Vyschnegradsky. They were played on four pianos, two tuned at concert pitch and the other two a quarter tone higher. At first, the music sounded like dialogue delicately off key. Then—to this listener, at least—the auditory experience became a surprising and liberated pleasure, comparable to the experience of hearing Debussy's whole-tone-scale music for the first time. It was nothing like hearing the stiffer diatonic music or the work of any of the twelve-tone composers. Vyschnegradsky's quarter-tone music is iridescent, savory, and richly tactile—a sort of supersensory production, which, though received through the ears, involves the other senses. Being a music of fractions, it is refined, calculated, curiously unequivocal, and balanced. It seems there was a Paris concert of it in 1937, when one of the critics referred to the composer's "attractive dissonant relationships between a tonic and its half-sharped fourths." Quite so. Rachmaninoff, Honegger, and Messiaen were early admirers of Vyschnegradsky. His "Ainsi Parlait Zarathoustra," performed by the Brussels Philharmonic Society after the Second World War, is considered his major work. Now living in Paris, he composes on a specially constructed, two-tiered, quarter-tone piano. He was born in St. Petersburg, studied harmony with a pupil of Rimski-Korsakov, and is the son of a director of the Imperial Russian Orchestra. Today, Vyschnegradsky is an invalid, as a result of his confinement in the Nazi prison camp at Compiègne. It is said that Stokowski will present his "Fugue No. 1" in America this coming year.

1952

January 2

The year ended with a significant clash between a couple of leading churchly and laic literary minds, a clash that has already become a *cause célèbre*. It consisted of an exchange of open letters between men who had been friends until the ink started to flow—François Mauriac, the noted ultra-Catholic novelist and editorialist of the daily *Figaro,* and Jean Cocteau, the noted, frequently pagan poet. On the surface, the quarrel centered on "Bacchus," Cocteau's new play, lately given a *Tout-Paris* opening by the Jean-Louis Barrault repertory troupe at the Théâtre Marigny, which Mauriac walked out on, offended by what he considered its frivolous sacrileges. Like Jean-Paul Sartre's recent "Le Diable et le Bon Dieu," Cocteau's piece, despite its mythological name, deals with the Martin Luther period in Germany and the so-called heresy of the new, free, Renaissance European mind, boldly seeking the liberty of personal truth. The two plays are also alike in that neither is its author's best. Cocteau's is adroit and stimulating; Sartre's is profound and, except for its splendid first act, boring. Mauriac's letter, which started the trouble, was run on the front page of the weekly *Figaro Littéraire* and looked like an encyclical, except for some relieving, very human and malignant opening touches, such as "Thou art at once the hardest and most fragile of creatures. [Both letters of rupture were, oddly, written in the second person singular, usually consecrated to expressions of love.] Thy hardness is that of an insect, thou hast its resistant shell. Nevertheless, if one squeezed a little too much—But no, that I shall not do." More loftily, he continued, "At the Marigny the other night, I suffered for the real Cocteau, the invisible Cocteau whom God knows and loves." With this, Mauriac started on his real sermon, reprimanding Cocteau for having used in his play a buffoon bishop and a cynical cardinal to represent the ever-noble Holy Church, and for having willfully composed subversive statements. Mauriac then went into redemption, Hell, saints, and God—not as a drama critic writing to a dramatist in a literary weekly but as a sacerdotal authority giving warning from a Sabbatical pulpit.

Cocteau's answering letter, obviously hastily written, was in print a few hours later, in the afternoon daily *France-Soir*. His letter was called "Je t'Accuse," recalling Zola's famous denunciatory letter in the Dreyfus case. Each of Cocteau's nineteen paragraphs began "Je l'accuse . . ." "I accuse thee, If thou art a good Catholic, of being a bad Christian," he said. He also accused Mauriac of being uncultivated, since the subversive statements Mauriac charged him with composing were taken from the Thomist Jacques Maritain, from Napoleon, from a medieval Archbishop Elector of Mayence, and from followers of Calvin. "I accuse thee of seeing only the ignoble in our world, and of limiting nobility to another world, which escapes us as incomprehensible. . . . At our age [both have turned sixty], one is no longer beautiful, but one can have a beautiful soul. I accuse thee of not having cared for thy soul. Thou canst insult me again," Cocteau concluded, deftly turning the other cheek. "I shall not answer. Adieu."

January 30

It is the dark-skinned antique peoples on the south shore of the Mediterranean who have been making sanguinary news lately, instead of the pale men of Europe, as is usually the case. That the Egyptians dared dream of breaking off diplomatic relations with Britain was less of a shock to Parisians than that they had dared burn Shepheard's Hotel, a caravansary that had come to seem an impregnable outpost of European civilization, with security and deft native service. The bloody Egyptian crises, flaming in the wake of the last fortnight's uprisings, deaths, repressions, and general strike in the French protectorate of Tunisia, which, in turn, followed on the touch-and-go situation in Morocco last November, have certainly stirred up Paris opinion. One type of Parisian thinks that a good, firm Metropolitan French hand is needed with the Tunisian rebels, who want a constitution. But the majority of Paris thinks that previous French governments are partly responsible for today's violence, because of their dangerous dawdling over decisions that might have soothed the natives—though they might also have vexed the French settlers. It appears that over the years the hundred and fifty thousand settlers in Tunisia developed the notion (without legal foundation, as the new young French Premier, Edgar Faure, has courageously made clear—too late) that, as co-inhabitants, they should enjoy equal power with the three and a third million Tunisians in any home rule, even that concerned with strictly Tunisian affairs. As for the government's dealing with the Moroccan upheaval, His Majesty Mohammed V, Sultan of Morocco, who refers to himself with a capital letter in the first person plural, in 1950 wrote to the French requesting talks on a new *modus vivendi;* in 1951 he had to write again, stating, "And for these negotiations We are still waiting." More recently, when His Majesty Sidi Mohammed el-Amin Pasha Bey, officially titled Possessor of the Kingdom of Tunis—a cultivated old gentleman whose dress uniform includes red trousers, a waistcoat, and jewels—received Jean de Hautclocque, cousin of the late General Leclerc, as French Resident General of Tunis, he said candidly, "You are the fourth Resident I have received since the war. Instead of changing Residents, the French ought to change their policy."

Parisians have shown enormous interest in the personality of the rebel leader Habib Ben Ali Bourguiba—now detained—whom few had ever heard of before he roused all Tunisia. His photographs show a violent, flat-nosed peasant face, rather than the more familiar aquiline beak of the elegant sherifian type. Until his followers began being arrested last week, he had twenty thousand disciplined militants behind him, and nearly half a million declared supporters. Frenchmen who knew him when he was studying law and political science in Paris, in preparation for his career of rebellion, say he is a born agitator, an intellectual, and an autocratic idealist who aims at victory but will accept martyrdom. He was born of modest parents in the coast town of Monastir, is now forty-eight, has long been married to a Frenchwoman ten years his senior, and, at one time or another, has spent four years in prison for his political activities. He became president of the Neo-Destour Party (*"destour"* is the Arabic word for "constitution"), which is Western-minded, Wilsonian, Kemalist, and plebeian. It was formed in the early nineteen-thirties, after a split in the Vieux Destour, or Old Turban, Party, which remains Koranic, Pan-Islamic, Oriental-minded, and conservative. Both Destourian groups are bent on independence. Those who know Habib Bourguiba say his personal tragedy is that his fine French education turned him from a Mussulman into an agnostic who believes in no prophet. They say he harangues mobs better in French, with Voltairian citations, than in Arabic, with flowery Koran couplets. As an eager intellectual, he fatally achieved the metropolitan cerebrations of those against whom, as a Tunisian, he vainly rebels today.

On May 21, 1890, Vincent van Gogh came from Arles, where he had been in a lunatic asylum for a year, to the small gray hamlet of Auvers-sur-Oise, near the river town of Pontoise, outside Paris. His brother Theo had sent him there for possible healing by the village physician, Dr. Gachet. Gachet had known Cézanne when Cézanne painted there, and, being himself a painter, was familiar with the crazed souls of artists. Vincent approvingly wrote Theo on June 4th, "The Doctor seems certainly as sick and bewildered as you and me. I am working on his portrait, the head very blond, with a white cap . . . a blue coat, and a cobalt-blue background, leaning on a red table with . . . purple flowers. He is absolutely fanatical about this portrait." The Doctor was less bewildered than Vincent, for late in July van Gogh borrowed an innkeeper's gun—to shoot crows, he said—and killed himself. He was buried in the churchyard at Auvers-sur-Oise.

Dr. Gachet's son, M. Paul Gachet, recently gave the white-capped portrait of his father—and the white cap, too, along with some other touching souvenirs of van Gogh and Cézanne—to the Louvre, and the pathetic little collection has just been put on public view at the Museum of the Jeu de Paume. In addition to the portrait, the collection contains works by Sisley, Pissarro, Renoir, Monet, and their friends. There are sixteen pictures in all. The only truly great items are three van Goghs—a self-portrait he brought with him from Arles, the Gachet portrait, and a picture of the back of the little Gothic church in whose yard he was eventually to lie. This trio of pictures has been on view before, but the Doctor's white cap—in a glass case, along with van Gogh's bamboo pens, for drawing, and his last palette, on which the colors are all smeared in diagonals, typical of his brush strokes in his wheat fields (it is shown next to Cézanne's palette, smeared with curlicues and cones of color, like his bathers' shadows)—gives them their earthly, intimate setting. In the glass case, too, is the Doctor's death sketch, in charcoal, of van Gogh's face—a fine likeness, sad and peaceful. The Doctor also did a portrait of van Gogh in bronze, perhaps from memory, with the bearded head placed in profile in a kind of dish—very strange. Given their present connection with the Doctor and his past, the paintings seem to lack the garish, art-show quality of long-viewed public collections.

For nearly a hundred and fifty years, Benjamin Constant's novelette "Adolphe" has been regarded as unique in lesser French letters, because it is the first example in romantic French literature of the delicate tragedy of the lady pursuing the gentleman, and of the inability of the gentleman, rather than the lady, to say yes or no. "Adolphe" is no longer unique, owing to the recent discovery of another minor Constant chef-d'œuvre, apparently lost in a trunk for more than a hundred years. It has now been published as "Cécile," though it might as well have been called "Adolphe the Second." It is even shorter than "Adolphe," being unfinished, but still more powerful, in its taut, nerve-racking, polite way, because it features two ladies pursuing one gentleman—Constant himself. One of the ladies is his eventual second wife, the Hanoverian Countess Charlotte de Hardenberg, who is the Cécile of the booklet, and the other is that Bonapartist exile and general brilliant holy terror of Europe, Mme. de Staël, whom he disguises as a Mme. de Malbée. His Charlotte was an obedient Germanic angel, but Mme. de Staël once quarrelled with him all night and all the next day, and she postchaised after him over every frontier of the western Continent.

"Cécile," to a greater degree than "Adolphe" (both are true stories), is a rare record of the new manners of liberty after the Revolution, and of how they affected the domesticities and divorces that had just come into style among well-bred, well-educated Europeans, who, turning from the upholstered conventions of the past, stepped boldly out of doors into a whole dangerous new modern world—our modern world. It was evidently written around 1810 and then dropped and lost as inconsequential. It is supposed that it was preserved in one of Constant's many mislaid trunks scattered over Europe; one of them is known to have finally reached his widow in Switzerland. All that is sure is that the manuscript, of twenty finely written big double pages, turned up in some Constantiana that a present-day member of the family, Baron Rodolphe de Constant-Rebecque, recently deposited in the Bibliothèque de Lausanne. It would be reckless to say that "Cécile" is more fascinating than "Adolphe," but it is definitely more Constant.

June 3

The Congress for Cultural Freedom's month of international twentieth-century masterpieces, presented and mostly paid for by well intentioned wealthy Americans, has finally come to its end, accompanied by the same mixture of general criticism and individual satisfaction it started with. Cold or hot, Paris owes thanks to the festival's program director, Nicolas Nabokov, for "Wozzeck," which it had been waiting twenty-five years to hear and which, together with the vertiginous peak of pleasure furnished by Stravinsky's evening with his symphonies, was the signal modern musical event of all the postwar seasons here. The Congress's modern-art exposition was justly ticked off for its paucity of French Fauves, but nobody hallooed the inclusion of two rarely viewed Picasso Cubist portraits—one of Wilhelm Uhde and the other of a "Jeune Fille à la Mandoline"—from the London collection of Mr. and Mrs. Roland Penrose. Benjamin Britten's new Melville opera, "Billy Budd," proved tedious and tactless. It was sung in English, and the only phrase that got across was "Down with the French!," which roused an icy breeze of laughter. A revival of the Gertrude Stein-Virgil Thomson opera "Four Saints in Three Acts," vigorously conducted by its composer, enchanted and tickled those who knew that it was a cellophane rose is a rose. *Figaro's* critic compared its Negro voices to savory, unctuous colonial fruits but remarked apologetically that Thomson's music reminded him of Erik Satie's, "from which nobody could tell whether he had renounced richness or resigned himself to poverty."

As for the writers' conferences on such hard knots as "Diversity and Universality" and "The Future of Culture," they proved that few writers talk well and most talk too long. Perhaps in consciousness of this, William Faulkner, who received an ovation from the French, merely muttered some incoherent phrases about how American muscles and French brains could form the coming cultural world and sat down, small, reserved, and inarticulate behind his little

Southern mustache. It was regrettable that creative Anglo-Saxon writers like Katherine Anne Porter, W. H. Auden, Glenway Wescott, Louis MacNeice, Allen Tate, and even James T. Farrell should have crossed waters to have, with one exception, merely journalists as their French colleagues on the platform. The single creative French writer to appear was General de Gaulle's man André Malraux. He made what was undoubtedly the most exciting, excitable speech of the conference, feverishly kneading his hands, as is his platform habit, extinguishing his voice with passion and reviving it with gulps of water, and presenting an astonishing exposition of politico-aesthetics that seemed like fireworks shot from the head of a statue. He never mentioned the General.

This has been a momentous fortnight for European history, which has been changed as much as it can be by statesmen's ink. Six statesmen signed up under the Quai d'Orsay's famous patient clock for an integrated European Defense Community. Their work can still be partially erased by non-ratification in any of their parliaments. Many French fear the whole business because it sounds like naïve New World idealism. The best thing about it may be that it operates on two of the least spiritual levels—that of pooled coal and steel and that of a pooled international army. General Eisenhower, who seemed to believe in the idealism, will be missed in France. He was the most loved and respected American citizen to function in Paris since Benjamin Franklin.

June 17

During the June *grande saison,* the quaintest and quietest event is the annual Concours International des Roses Nouvelles, of which the forty-fourth has just been held, like the others, in the ravishing topiary Parc de Bagatelle, where a jury chose the most beautiful new rose among fifty-five candidates, thirty-five of them French. The gold medal went to Flambée, officially described as "dark, brilliant red, with fiery reflections," which was grown by the French breeder Charles Mallerin. Even more impressive in the opinion of some was his Oradour, of a dramatic dark blood color, also with fiery reflections, which was named for the tragic town whose population was burned alive in the town church by some extra-sadistic Germans. Ominous dark roses seem to be this year's fashion. The only rose that was so new it looked positively peculiar was a fascinating little item frankly named Incendie, which had a center like pink gas-log asbestos and outside petals of burnt-out lavender. A pupil of Mallerin's, Francis Meilland, whose Mme. A. Meilland, named for his mother—yellowish with pink edges on an extraordinarily strong stem—was the rose sensation of recent years, has exported it to the United States under the name Peace, and over two million bushes have reportedly been sold there, making its breeder a fortune. This year, Meilland got no more than an honorable mention, for a handsome ochre rose called simply No. 19-51.

The roses are judged on six characteristics, with two points for *floribondité,* three points for vigor of vegetation, three for resistance to maladies plus beauty of foliage, four for qualities of newness, and four for general character—in which perfume figures only slightly. This year's winners were a delight to the eye, not the nose. None of them smelled much. According to gossiping rose men at Bagatelle, Louis XIV's gardener knew only fourteen varieties of roses; today there are fifteen thousand, and at least two hundred new ones are produced annually, some by "somatic accidents." Aside from the whistle-blowing of the park's spirited policemen whenever a visitor openly sniffed a blossom or caressed a bud, the Bagatelle rose competition was an exquisite performance and furnished an enchanting scene—prize-seeking multicolored roses as high as fountains on trellises or arranged like hedges or loading down bushes, and all these masses of carefully arranged color within an ancient garden of boxwood cut into cones and high embrasures, with, in the near distance, on a knoll at the edge of the blue-hazed forest trees, a Chinese-looking gazebo, serving countrified refreshments.

July 1

The most fastuous spectacle ever offered in the present Opéra house, and one superior in rich costuming, at least, to many staged at the Folies-Bergère, was the June-season revival of Jean-Philippe Rameau's "Les Indes Galantes," first given in 1735 at Louis XV's Académie Royale de Musique. Unfortunately, ultrarefined Parisians think the Folies-Bergère is just where the new Opéra show belongs. With the *polloi,* it has been a sellout. A royal *ballet héroïque,* "Les Indes Galantes" consists of plots, solo singing, choruses, verse declamations, choreographic *divertimenti,* a fable, and a moral. According to the fable, a European love feast, held in what looks like a bit of Versailles on Mount Olympus, is interrupted by a call to war. The moral is provided by a flock of cupids, danced by the *rats de ballet,* or the little girl ballet beginners, who take off, on an escalator rigged as a sunbeam in the Olympian sky, for climes where Venus is still supreme—specifically, Turkey, Peru, Persia, and an unnamed blackamoor isle. To these loving lands, the audience and practically the entire Opéra troupe go with them. This gives you the idea.

It is the overgenerosity of the production of "Les Indes Galantes" that makes it irresistible, with the Opéra's huge stage opened to its back wall for the ensembles, and crammed with ballets, marching choruses, star Opéra singers, a kaleidoscope of exotic costumes, and trick scenic effects that include a goddess floating on a cloud, ballet notables popping up through trapdoors behind disappearing gardens, perfume sprayed over the spectators to accompany a *divertimento* called "Les Fleurs," a shipwreck, a sweet-breathed Peruvian volcano erupting real fire, smoke, and incense, and an earthquake with falling temples. The Peruvian scene—which has a Lifar ballet of masked dancers costumed in clashing purples in a modernist abstract manner while preserving the Louis XV sartorial style that marks the whole show—is, artistically, the peak. There is also the sight of Mme. Geori Boué, usually seen as a seductive Thaïs, singing the Savage Queen in what looks like mammy blackface.

July 15

The schism in General de Gaulle's Rassemblement du Peuple Français is regarded as having been the most sensational event of this year's first Parliamentary term. It furnished the kind of intellectual political hubbub the French esteem and enjoy, for it ranged from mysticism to Olympianism and included rarefied theories of legitimism, pragmatism, and patriotism, plus the philosophic struggle between doctrine and action. In plain terms, it was really an old-fashioned bustup between the common sense of the Party militants and the purity complex of the statuesque Party leader. It was the sort of split that periodically splinters all intelligentsia-led European political groups. It really began in May, when so many R.P.F. deputies were upholding Premier Pinay in the Assembly, contrary to the General's strict orders, that he composed a stiff form letter of reproof, which began *"Mon Cher Député—,"* with the names of the couple of dozen deputies typed in as needed. On being told last week, at the Party caucus in St.-Maur, that they must in future vote in Parliament the way the General, who is not a member of Parliament, orders from the sidelines, twenty-eight dissident deputies resigned from his Rassemblement and overnight organized themselves into a brand-new wilderness party of the Right, which they named Groupe Indépendant d'Action Républi caine et Sociale. In the suburb of Ivry, four R.P.F. municipal counsellors, including one woman, who was vice-president of the R.P.F. National Committee, also resigned, presumably to found a suburban wilderness party. Then a handful of R.P.F. senators resigned. Since French senators and deputies often double as mayors of their home towns, de Gaulle, in losing about forty of them, also lost some important mayoralties. The General's loss of face and Parliamentary followers is large, ironic, and probably not fatal. His R.P.F. deputation, which was the biggest in the Assembly, has merely shrunk itself out of first place.

According to Deputy Edmond Barrachin, chief of the de Gaulliste dissidents, the rebels are tired of voting no to everything, which has been the General's policy for his Rassemblement ever since 1946, when he resigned as chief of France. They are tired of being in opposition to every government and thus "voting more often than not with the Communists," the other perpetual Fourth Republic opposition. To serve France, the defrocked de Gaullistes are eager to go along with any party that shares some of their ideas, in the end hoping to dominate it from the inside. They are weary of the General's strange salvational conviction that only when everything else in France fails will the desperate country give him—him alone—a loud call. Barrachin said, "To wait in immobility for the national catastrophe, assuredly without wishing for it, so that General de Gaulle can be called to power seems to us an insufficient plan of action."

In the past week, the French and Europeans and their newspapers have been full of talk about de Gaulle. No two comments on him are alike. That he is still a magnetic crusading figure, able to attract men, whom he then puts to the exhausting task of waiting and doing nothing until history happens to him once again, is proved by the fact that his Rassemblement swelled to the biggest single party in France at the last elections. It was corroborated last week by the combined malice and satisfaction his hosts of enemies expressed at the news of his Party's schism, thinking that it weakens his chance of a coup d'état, which he says he disapproves of, too. Some of the things de Gaulle wants today are wanted by many other Frenchmen, who certainly do not want him also: a revised constitution, rechannelled patriotism, a centralized government carried on exclusively by men who were in the Resistance during the war and who would form a new aristocratic governing class, founded on bravery, and what he is always talking about, "the great social and national highroad that will lead to a regrouping" of France's welter of parties—a kind of upper-class Left of social reform, which still sounds like Fascism to the European working-class Left.

August 12

France is suffering from the worst epizootic of foot-and-mouth disease since 1939, and one of its most depressing aspects is that it is being blamed on the Americans, in an agricultural version of the Korean War microbe propaganda. The contagion has also increased Premier Pinay's unpopularity with the farmers, who think his government should have been ready with adequate vaccination virus. This is the first time that a French government was supposed to foresee an epizootic. Though the Pinay government had nothing to do with it, a vaccine that has been under study for the past five years at the Institut du Radium and at the Laboratoire Central de Recherches Vétérinaires was put in quantity production last week. It is dry, and cheaper and stronger than that previously imported from Switzerland for general use here; where one cow could provide only a few grams of the old vaccine, she can provide ten to fifteen kilos of the new stuff. If the new formula does its work, it will give French laboratories considerable honor and European farmers and cows great relief.

Since American farmers are thought by French Communist farmers to be the monsters who sent over the epizootic, it might interest our dairymen to know that this onslaught is produced by a new, stronger type of virus, never seen here or in America before;

that it started in Denmark last year; that the French farmers don't destroy their ill animals but merely report them to their mayor, stick up warning cards on their farm gates, and feed the sick animals grain mash or mashed potatoes or anything mushy that won't hurt their sore mouths; and that seventy-five to ninety-five per cent of the French cows recover. This must seem hardly cricket to the English, who are manfully destroying their sick animals during the present plague, which is also rampant in Belgium and Holland.

October 20

Until last week, it looked as if the Paris newspaper reports on the American Presidential campaign were going to consist of travelogues on the United States. For instance, there was the *Figaro* reporter's account of the fine time he had in the Eisenhower press plane, flying for two days around nothing but Texas, which he had never seen before and which he vastly enjoyed, taking particular pleasure in a mounted cowboy on some Republican greeting committee who told the Frenchman that he had never been on a horse before, being used to riding herd in a helicopter. At the end of the *Figaro* travel report, and well after the cowboy, General Eisenhower got a mention: "The crowds that came to hear Eisenhower were not large for this state of superlatives." Then, there was *France-Soir's* reporter, who had himself a splendid time at the New Orleans welcome to the General, which he wrote about as though it had been a Nice Mardi Gras carnival. Considering the offense taken here in France at Eisenhower's convention reference to the French as morally debilitated and half agnostic or atheistic, it is extraordinary how the French reporters allotted him the major publicity. Now they are also allotting him the major criticisms.

At first, the unknown Governor Stevenson was literally never mentioned in the leading Paris papers. Even his being *terre inconnue* was not treated as news. The opening pen portrait of him was given in *France-Soir* by its reporter, who saw him in Milwaukee: "He is a man of medium height, pink face, intelligent eyes, and Roman nose, possessed of elegance of manner and language, and yet capable of drolleries that make the people laugh. His speech here was on the academic side. It proved that he is a marvellous orator, far superior to Ike." Then the superior orator was dropped back into limbo. This past week, he has been brought forth again, with praise of a type which indicates that the visiting French reporters, at least, may have made their final personal selection. He is now suddenly hailed as "the great discovery from the West," the new midland American with *des éléments de grandeur dans sa nature.* A potpourri of their comment runs: "Stevenson's chances are augmenting"; "Stevenson appeals to the voter's mind, Eisenhower to the voter's emotions"; "Stevenson is a personality a hundred times more complex than his adversary" (high praise from a Frenchman); "Stevenson's speeches show his full stature as a statesman"; "If he wins the election, it will prove that the American voter has thought things over carefully, despite the organized theatricalities that are the *mise en scène* for the campaign, and maybe even despite television." The *Monde* reporter's latest comment on the possible outcome of the election is characteristically grave. He points out that it is not without irony that Negro leaders have said the next President may owe his victory to the colored people, "the so-called undesirable tenth [of the American population], who savor in silence this unexpected revenge after so many years of humiliation."

The French will certainly breathe easier when the election is over and they know whom they are going to get. They will be particularly relieved to have the campaign over, and an end to the candidates' competitive denunciations of Communists, Premier Stalin, and Soviet imperialism, which sound, some French think, too much like the Communist daily *Humanité's* denunciations of Wall Street varlets, President Truman, and Yankee imperialism, and make them nervous, placed as they are by history and geography as a way-station battlefield and occupation ground for whichever of the two twentieth-century giants should tire of being insulted.

Game birds now in season seem relatively cheap, considering how expensive even a pork chop is. Because game is extra-plentiful this year, fine pheasant and quail cost about the same as last autumn, which gives them the fallacious appetizing taste of a bargain. Even ortolans, those luscious, musical little mouthfuls, are available. They are seasonal here twice yearly, with a few in February—mostly lost last winter in high winds—and more in September and October. Ortolans are one of France's smallest birds of passage. They are brownish-grayish, gregarious buntings; but the peasants have always affectionately called them sparrows. The flocks, which mainly reside around Les Landes, in the southwest, fly off to summer in East Prussia and return in early fall to peck in the vineyards and glean in the grainfields, which is where they are hunted. The method of their capture and the subsequent stuffing of them to the limit of corpulence within their small measure has not changed since they figured as dainties on royal menus centuries ago. To catch them at this time of year, nets are set upright in the grainfield stubble, and live ortolan decoys are tied to them, to flutter invitingly. The hunters blow on little whistles devised to imitate the ortolan song, which is pretty. The birds, flying in dense coveys, must be entangled in the nets by the hundreds to make the hunt worth while. The hundreds are then packed into cages, are placed by the peasants in dark rooms, to keep them drowsy and inert, and are stuffed in the gloom for a fortnight with millet seed. Legend says that table ortolans meet their end by being strangled by the peasants with a horsehair. In reality, the peasants do it with thumb and forefinger. By that time, the ortolan is so plump that the back end of his little carcass is almost square with yellow fat. "He is so anemic—that is, bloodless—with fat as to be comatose and unconscious of death," a Paris chef classically said, and he added briskly that anyhow it was the natural destiny of ortolans to be eaten. Not many Paris restaurants specialize in them any more. This season, they cost two hundred fifty francs each, and two—or at most three—make a serving, for they are the richest bird in the gourmet's aviary. They are not supposed to be touched by knife or fork (though honestly they taste fine on a fork); the eater is supposed to seize them by the beak with the fingers of one hand and consume them, bones and all, beginning with the feet. The diner's free hand should be used like a cover over the ortolan to capture its fragrance, which is enticing; the bird's role is to please both the olfactory and the gustatory sense of man. It is ritually chewed slowly, to give it a chance to melt in the eater's mouth. Old engravings show ortolan-eaters with napkins hoisted like tents over their heads to enclose the perfume, and maybe also to hide their shame.

November 5

The Presidential election felt more local to the French than any other in American history. They were on the *qui vive* to the end, to know who was winning, who had won. By

Tuesday noon, Parisians had begun listening to their radios as if they could already hear the early-morning votes dropping into the ballot boxes across the Atlantic. And when the Wednesday news of Eisenhower's victory came in, they reacted to it, most of them, with chagrin. After all, Stevenson had, in a matter of weeks, become their man—their unspoken choice, who they had hoped would carry off the prize in a final sweep.

As the President still occupying our White House must be aware, there has been a lot of worry in Washington and Paris over something that the State Department and the Quai d'Orsay do not mention out loud officially but that the unofficial French have talked and written about incessantly for the past two years—French anti-Americanism. The main reasons for it have just been conveniently catalogued by the weekly *Paris-Match,* a French imitation of *Life,* which first, the week before, politely listed the reasons the Americans don't like the French. Americans living in France—provided they have ever met any French, that is—know both lists by heart. A list of reasons for the French anti-Americanism that has now become so openly important runs about like this: Basically, the French, after seven years of peace, are tired of being occupied, even if for their own welfare, and especially of being occupied by Americans who, in their own phrase, "never had it so good." There are several thousand Americans in Paris alone, working in the intricate Washington services here, and thousands more, including the military, elsewhere in France. Since human beings are social animals, the first French anti-Americanism is cruelly social: The French think too many Americans talk and drink too much and too loudly in public. To jump to the more important economic realm, American high tariffs and high ideals, mixed, are a major anti-American canker. According to the French, what France wants now is trade, not aid, from America—business with America, not noble charity from it—no matter how much France has enjoyed or still needs aid. While deeply grateful for the Marshall Plan, the French say that the most intelligent thing about it was the Americans' original time limit on it, indicating a wise realization as to how much giver and receiver could psychologically stand. A commonly expressed opinion of many French and some West Europeans is that economic aid no longer provides the desired easy, early solution. The aid the French do want is military. Their six-year-old war in Indo-China has, according to

President Auriol, cost them nearly double the American aid they have received for it, is three times as old a war as the United Nations-American war in Korea, at the last accounting had killed several thousand more Frenchmen than the Korean war had killed Americans, and is even more unpopular, many times over, with French youths and their parents. (American contempt for France as a military ally is another sore source of anti-Americanism.)

The next reason for anti-Americanism here is certainly our high immigration walls, now guarded by a Gorgon Statue of Liberty, which the French say they are beginning to think they misnamed when they gave it to us. That a French student with a year's scholarship in America must, in obtaining his United States visa, not only furnish the American consular authorities here with his complete young life history but also subject himself to a medical examination of an embarrassingly inquisitive sort seems to the natives of France, which has for centuries been a haven to all, including sick geniuses in exile, a ludicrous example of our new, strangling red tape. Intrusive visa questionnaires and visa difficulties generally—especially for professors or any other kind of French brains—have soured our reputation for hospitality.

In international affairs, far and away the bitterest anti-Americanism still springs from our rigid anti-colonialism. The French respect our historical loyalty to the principle as decendants of a successfully rebellious colony in 1776, which quite properly possesses no colonies in 1952. But in our insistence that everybody else should be like us, the French say, we have helped ruin postwar Holland and England—countries we had fought to save—by encouraging the Dutch loss of Indonesia and by cheering on the breakup of the British Empire, which, over the years, controlled the very Mediterranean and Near and Far Eastern lands—even those unconnected with the Crown—that so worry us now: Egypt, Persia, India, and, of course, China. Wise to the delicate Dixiecrat issue in our Presidential election, anti-American Parisians ask how about our concentrating on our treatment of the colonials within our own home, i.e., our Negroes?

Next to the last of the major causes of anti-Americanism here is what the French jeer at as the irony of our hysterical hunts for a few hidden Communists, "like moths in an office carpet." Obviously, the final cause of anti-Americanism is the French fear that, after having liberated France in one war, in our clumsiness we might plunge her into another. *Paris-Match* summed up in part: "No country was worse prepared than America for the worldwide role that she brusquely had to play, and on so immense a scale. The Americans are leaders through the force of events, but without having the desirable first-rate qualities like sure judgment and a cool head. . . . They still lack almost completely the realism that the old imperial nations, like England, Holland, and France, learned from their own animated histories."

November 19

France has just lost two once important public figures, who could not have been more unlike—the harsh old royalist Charles Maurras, aged eighty-four, and the limpid Surrealist Leftist poet Paul Eluard, still in middle life. In January, 1945, a handful of us journalists managed to travel to Lyon, where Maurras was being tried for collaboration and was convicted of treason against the Third Republic. The trial was held in a bombed-out courtroom with snow filtering through its broken windows, near which Maurras sat in the prisoner's box, holding his walking stick (though he was going nowhere except to prison for life) and with a huge muffler, twisted like a fatal noose, below his large, useless ears. He was doubly deaf; physically he could not hear, and mentally, through habit, he was unable to listen to reality. Born in Provence and later, in Paris, a young protégé of Anatole France, he was by profession a self-constructed paradox. He began as an anti-Dreyfusard (who thought Dreyfus's final shame was his innocence), became a royalist (because he did not like republics) and then a holy monarchist (whom the Vatican put on the Index and the French royal pretender disavowed as a nuisance), was anti-British, anti-Semitic, anti-German (though he collaborated with Herr Abetz and was pro-Pétain), pro-Fascist, and forever pro-French. He founded the royalist Ligue d'Action Française, inspired Les Camelots du Roi, or King's Peddlers, who used to riot in St.-Germain-des-Prés, and, though by literary gift a purist, turned out, with Léon Daudet, daily virulent, scabrous editorials in their royalist gazette, *L'Action Française.* Both went to prison briefly for provoking a murderous Camelot attack on Léon Blum, who Maurras had written "deserved a carving knife." Even before this last war, Maurras' historical paradoxes began falling apart. At his Lyon trial, he already looked like a captured malicious, intelligent mouse—all beady eyes, swollen nose, vast ears, and diminished body. In prison these last years, he still scribbled royalist propaganda. He reportedly died lucid, writing up to the end.

Paul Eluard, whose real name was Eugène Grindel, was born in the Paris suburbs but early fled to the climate of the poetry of Rimbaud, Lautréamont, and Apollinaire, and then, after being hideously gassed in the First World War, to the new insurgent poets, and to the Dada movement, in the company of Tristan Tzara, Louis Aragon, and the painter Picabia, who later became the renovating, influential Surrealists. Eluard was a familiar figure at the Deux Magots, their café headquarters, and, unlike them, was a serene man, with a gentle, handsome, medieval face. In 1924, he disappeared and was thought dead, but he had merely quietly embarked on a tour around the world. On his return, his book of poems "Capitale de la Douleur" and the lyric phrase in one of them, *"Je chante pour chanter,"* became famous. During the last war, despite fragile health, he joined the Resistance, and emerged formally engaged with its Communist groups. His poetical material remained always private and poetic—love of love, of visionary liberty. In one of his vers libres, "Liberté," appeared his famous lines:

> *Sur toutes les pages lues*
> *Sur toutes les pages blanches*
> *Pierre sang ou cendre*
> *J'écris ton nom.*

As even the conservative *Figaro* has just said, "Communist or not, Eluard's lines on liberty should be known by every child in France."

December 2

France's war in Indo-China has been going badly, even tragically, with the ever-stronger Vietminh forces encircling French military outposts. The fighting is now so clouded by Army censorship that soldiers' parents here can only fear the worst. General Eisenhower's projected visit to the United Nations war in Korea, with its recent horrible bloody seesawing, has filled the French with the hope that first one, then another truce, on whatever terms, will bring a cease-fire for the New Year.

As for Pinay's government in France, it has been touch and go. For days it has looked in the evening as if it would fall, but it is always still there in the morning, like the not very radiant winter sun.

December 30

In one way, anyhow, French politics stuck to a pattern in 1952. In the first week of the year France lost her government, and in the last week of the year France lacked a government once more. It has lately looked as if France would have no government for New Year's Day of 1953, either. The new government, when it does turn up, will be the eighteenth—a dozen and a half of them—since the postwar Fourth Republic was founded, in 1945. The present hiatus is regarded by the French as the most profoundly serious of all, if only because it proves for the seventeenth time that something is wrong with the system, which it is now admitted must be fixed. Until recently, the demand for constitutional revision was suspiciously considered exclusively the product of the ideological mysticism of General Charles de Gaulle and his followers in the Rassemblement du Peuple Français. Today, few politicians would disagree with the General—not even those who wrote the Fourth Republic's constitution. The main trouble with the document is that it makes better provision for the fall of governments than it does for the coördination of their limited postwar powers.

It is significant and astonishing that after Premier Antoine Pinay's fall last week, de Gaulle's Rassemblement was given—if only in vain—its first chance to form a government. It was too much to expect that the "Trojan Horse" (which is what jeering anti-Gaullists, in disapprobation and fear, call the General's Party) would be able to take the preliminary jumps, let alone stay the course, loaded as it is with the weight of his remarkable personality and the personalities of those devoted to him. But that the Gaullists were asked at all showed a new Parliamentary attitude toward Gaullism, and certainly showed a new attitude on the part of the Gaullists toward themselves. Previously, their General had declared that his group should not even aim at taking power unless he could be the governmental chief, preferably President of the Republic; today he is not even a deputy—is, indeed, nobody but the fantastic head of his own Party, living in secluded campestral concentration in his country house southeast of Paris. He also instructed his deputies to vote nothing but no in Parliament, where they sat like monuments of obstruction until recently, when thirty-one of them revolted and were thrown out of the Party as rebels. In accepting an offer to try to form a government last week, the Gaullists, on the General's orders, emerged at last from their mysterious isolation, really with the object of reviving the faith of their voters, who were tired of the General's dignified policy of patiently waiting for France to collapse and then summon him as savior. What the other politicians hope they have gained from letting the Gaullists have a nibble at reality and power is that the Rassemblement will now permanently abandon its position of negation and help form a more stable coalition majority, without which no party is strong enough to found and hold a government in France today.

Jacques Soustelle, the Gaullist deputy who failed to become the new premier, is one of the group's several noted intellectuals, having been a distinguished ethnologist and the assistant director of the Musée de l'Homme before he took up a political career. As one accustomed to dealing with various species of man, he expertly interviewed ten former French premiers, of ten party mixtures, all of whose short-lived governments had fallen like houses of cards. Premier Pinay, in his bitterness over his recent fall, angrily described the French Parliament as a "bear garden" and declared that he would take part in no more governments. Experts opine that before long he will probably be called to take part in a government of his own again. The Banque de France warned last March that the state was in danger of bankruptcy, and it was the theretofore unknown Pinay who came in as a little businessman to save the franc. In the one week since he left, the franc has fallen and the dollar has risen. Financially as well as politically, Pinay may be the only premier France can now afford.

The Bibliothèque Nationale is holding a superb exhibition to commemorate the fiftieth anniversary of Emile Zola's death. It is the kind of amalgam of scholarship, literary odds and ends, leftover household objects, and revived atmosphere of an epoch that the French project so well, especially out of the nineteenth century, where they have such sure literary wealth to draw from. With characteristic thoroughness, the show starts with Zola's grandfather's birth certificate, issued in Venice, where the Zolas originated, and ends with a cast of the writer's hand in death and a copy of the discourse delivered at his tomb by his friend Anatole France. To set the background of Zola's greatest sociological novels—the series about those degenerating families, the rich Rougons and the impoverished Macquarts—a hideous, opulent Victorian parlor, suitable for the rich branch, has been assembled, as a sort of illustration. Then comes Zola's hand-written manuscript of "La Fortune des Rougon," with the solemn opening lines of its preface—"I want to explain how a family, a little group of beings, moves through society . . . giving birth to ten, to twenty individuals. . . . Heredity has its laws like those of gravity"—which laid the basis for this first naturalistic French novel on social evils. The rarest items are early paintings by Cézanne, friend of Zola's youth at Aix, which include a self-portrait of 1858 that looks like a Greco, and some fumbling young landscapes of the Aix countryside, then newly transformed by the famous dam built by Zola senior, an engineer. There is also shown a wealth of pictures of and by, and letters to and from, the great painters who made up the later Paris epoch Zola plunged into—Manet, Monet, Daumier, Toulouse-Lautrec, and van Gogh (with a still life of two books, one Zola's best-seller "La Joie de Vivre," the other the Bible). There is correspondence of the literary gods Sainte-Beuve, Flaubert, and de Maupassant (with a letter to George Sand complaining that she had written no puff articles on Zola's newest book), and there are stately communications from the Goncourt brothers. However, the dramatic, vital part of the show is the documentation on Zola's participation in the Dreyfus case. Here is shown his hastily scribbled, bold manuscript of "J'Accuse" and his printed copy of it—laconically marked by him in pencil *"A Garder"* ("To be saved")—as it appeared in Clemenceau's newspaper *L'Aurore,* where it shook Paris and Europe. There is also the French *état-major's* original photograph of the crux of the Dreyfus case, the notorious Esterhazy *bordereau.* The exposition catalogue states that the *bordereau* itself was, ironically, lost in 1940 during the exodus from Paris, after which the anti-Dreyfusard Marshal Pétain assumed anti-democratic power in France. The exposition is a thrilling historical treat.

There are not many living major French poets today—really barely enough for the literary circles to squabble about. Least contested as France's dominant poet of philosophy is René Char. He is a forty-five-year-old country giant, brought up among peasants in the Vaucluse, beside the river Sorgue, which wanders liquidly through his strophes; earns his living by manufacturing concrete building bricks; and was a redoubtable military chief of the Maquis in the war, when he kept a now famous militant diary, "Feuillets d'Hypnos," which, in exalted prose, records death, hope, and reality as the French knew them in the *forêts*. He is a man of limited schooling and vast reading; the main influences on him were the pre-Socratic Heraclitus, and Nietzsche and Hölderlin. His poetry, stemming from Dadaism and briefly Surrealist, is, for one thing, the opposite of T. S. Eliot's, being a sort of triumphant cry from classic individual man on the rack of the modern machine age, which his voice rises above. Char was recognized as early as 1930. His major postwar works are the volumes called "Fureur et Mystère" and "Les Matinaux," and his most recent one, "A une Sérénité Crispée." In his occasional love poems, he seems to French ears to achieve again the sonority of the senses that they heard in Mallarmé. Most of Char's poetry is in prose, consisting of aphorisms that link nature—his river, the sky, and harvests—to an aspirational philosophy so condensed as to read like an inscription on stone.

Char's influence on the non-Existentialist young French is enormous, though he is little known to Americans and English. There has been considerable animation in poetry circles here on seeing the first big selection of Char's work for foreign consumption—thirty-five pages, half in French, half in spirited English translation by Professor Jackson Mathews, of Washington State University, which are featured in the autumn issue of *Botteghe Oscure,* published in Rome, which has just arrived here. It is still the leading European literary review and probably the only important one in four languages—Italian, French, English, and American—addressed to the literati of Rome, Paris, London, and New York. In the present number, there are fewer Italian contributions than usual. There is a lot of English poetry, since the American-born Princess Marguerite Caetani, of Massachusetts, who is editor-publisher, is an optimistic believer in a renaissance right now of fine lyric English poets, led principally by Dylan Thomas. To the Old Guard French literary people who recall the Princess Caetani's Paris-published literary magazine *Commerce,* back in the days of her publishing association with Paul Valéry and André Gide in their earlier glories, her Rome production seems a distinguished lineal descendant.

1953

January 14

The incoming French government has taken on three faces, though one is usually all that can be managed. There is the regular face presented by the Fourth Republic's eighteenth government, just set up by Premier René Mayer—a Left and Right of Center compilation, whose principal local tasks are to revise the Constitution and to keep the franc, and itself, from falling. There is also the new governmental face presented by the Gaullists, hitherto hidden by their mystic veil of parliamentary nonparticipation but now openly and unexpectedly influential as the dominant coalition support. Behind them loom General de Gaulle's strict, proud views favoring French sovereignty; his ideas against the European defense community in its present aspect and against the international European Army as now planned; and his warm convictions against a Federated Europe, which he wishes to see only as an unfederated political Western union. As insurance against these last three policies, he and his party demanded, and received on a platter, the head of Foreign Minister Robert Schuman, the old Lorrainer who, like a kind of borderland godfather, was officially responsible for them. Yet it is precisely the idea of Federated Europe, the vastest idea of the three, that has just been amazingly accelerated in the final votes of the *ad-hoc* meeting of the European Constitutional Assembly, in Strasbourg. There, on a historic occasion, representatives of the six member countries' parliaments—with, from France, sharp Independent Republican Paul Reynaud, the Socialist Guy Mollet, and the Gaullist senator Michel Debré—met to take a decisive, if argumentative, step forward toward a functioning European assembly for "a political community, supernational and indissoluble, founded on the union of the people." And it is specifically as the creative intelligence behind these modern ideas that France shows its third face—the face of Europe's leader in political thought, of the revived France, whose imagination produced the only new European concepts offered since the war.

The confusing irony behind this portrait—an irony to which the French are highly sensitive—lies in the fact that quantities of French citizens, including members of Parliament, are increasingly dead set against these ideas (though not against the more materialistic one that produced the Schuman coal-and-steel pool, already functioning at a great rate), and especially against the ideas of a Federated Europe and a European Army, with its inclusion of a rearmed Germany, France's traditional enemy. Regarded as attractive theories left over from the postwar past, when men's imaginative energies were freer, they are now deemed, with growing conservatism, impracticable for the present and better reserved for some undated tomorrow.

It looks as if Premier Mayer were already being squeezed between two generals—the French one and the American one who is about to be President of the United States. It was to General de Gaulle that Mayer made his pre-Premier promise of so-called annexes, or changes, in the European Army setup, in return for help in forming a government. It is on Eisenhower that, as head of our government, Mayer plans to call soon, supposedly in regard to aid for France in her military commitments in the Indo-Chinese war.

Last year was a difficult time for getting down to brass tacks on Franco-American understandings, because the United States was having its elections, and this year may not be much better, because now it is France's turn at the polls. It is having municipal elections in the spring, which often cause governments to fall, and in the early winter it has the election of the President of the Republic, when, traditionally, the Cabinet resigns, taking the government with it. So whatever Premier Mayer is up against, he may not be up against it long.

January 27

For the past three weeks, a prolonged fog has transformed Paris into a phantom capital of gray beauty, of which everybody is growing extremely tired. The anticyclone has covered all France, and so has *la grippe,* in one of the greatest epidemics since the plague of Spanish influenza in 1918. Although the current visitation is fortunately only a mild form of the virus, it has laid the inhabitants low, especially in the cities and countrysides of Clermont-Ferrand, Bordeaux, Rennes, Strasbourg, Amiens, and Paris. A thousand Paris telephone girls are reported sick and off the job, along with thirteen per cent of the saleswomen in the big stores and thirty-five per cent of the Palais de Justice judges (but only twenty-five per cent of the lawyers, who seem tougher).

The persistent fog has acted like a melancholy condenser of people's present apprehensions. The French are startled and worried by the news—coming so soon after the war's deadly ambitions and racist horrors—of the Nazi revival and the arrest of its presumed leaders in the new, republican heart of their traditional despoiler, Germany; by the revival of anti-Semitism, with its arrests and fear, in so many places; and, not least, by their own republic's governmental instability. And, to make it all worse, there are France's deteriorating, tit-for-tat relations with the United States. After what Parisians have so far viewed of 1953, they will be glad when the opening winter segment is passed and the fog has lifted and they can see the spring, with the world still in place.

A sharp modern genre novel, "Au Bon Beurre," by Jean Dutourd, which last month was awarded the Prix Interallié, is now a best-seller. Enough time has elapsed since the war to allow Dutourd's characters and their cruel, comical avarice to arouse laughter, in retrospect. The family he features—a M. and Mme. Poissonard and their mealy little son and delectable daughter—are already classics, personifying those Paris shopkeepers in the butter, egg, and cheese trade during the Nazi Occupation who were known as *bofs,* the initials of the *beurre, œufs,* and *fromages* on which their dishonest fortunes battened while their customers grew thin. It is a matter of history that the *bofs* emerged as the *nouveau-riche* phenomenon of Paris, where they still climb. In "Au Bon Beurre," the Poissonards begin their climb by watering their milk and praising Marshal Pétain's doctrine of suffering. Their prosperity eventually leads them to Vichy to present duck eggs to the Marshal, which furnishes one of the driest, funniest tableaux in the book. Climbing ever higher on black-market produce, on hypocrisy, on cruelty, and on clichés that nearly prevent even them from knowing what monsters they have become, they step from Nazism into Resistance Communism; into democracy, as the Americans help liberate Paris; and then into prison as collaborationists, until they buy their way out. In corroboration of their *bof* belief that a big enough fortune always justifies the means, their daughter is then sought in marriage by a sensible poor young nobleman, who truly loves her, her parents' money, and his hope for a new democratic postwar France. Dutourd's scenes are scraped right out of real life, and his dialogues furnish the wonderfully accurate accompaniment.

As a secondary accomplishment, Dutourd has done the French translation of Hemingway's "Le Vieil Homme et La Mer." It is one of the greatest successes Hemingway has ever had in France. Even a costly de-luxe edition on fine paper is already out of print. In an amazing appreciation, the writer Jean Guéhenno stated that he was putting it on his library shelf beside "Feuilles d'Herbe," by Whitman; "Un Cœur Simple," by Flaubert; "Des Souris et des Hommes," by Steinbeck; and "La Mort d'Ivan Ilitch," by Tolstoy.

To honor the eightieth birthday of France's writing genius Mme. Sidonie Gabrielle Colette, born in St.-Sauveur-en-Puisaye, Yonne, on January 28, 1873, and still writing today in her apartment in the Palais Royal, literary weeklies here all fêted her with photographs of herself and with the compliments of others that have accrued over her lifetime. Facsimiles of letters addressed to her were printed, among them a typical one from André Gide presenting "praise you certainly did not expect. I, too, am astonished at my writing to you, and at the great pleasure I had in reading you. I devoured 'Chéri' in one swallow. Already I wish to read it again, but I am afraid. What if on rereading I find it less good?" Marcel Proust confided, in his wavering, evasive handwriting, "I wept a little this evening on reading 'Mitsou.' Humbly I compare your restaurant scene with those of my Swann, whom you do not yet know." Claudel, Valéry, Francis Carco, Jacques de Lacretelle, André Dunoyer de Segonzac, and Pierre Fresnay, along with Rebecca West, Katherine Anne Porter, and Rosamond Lehmann, contributed praise from near and far. Intimate letters of fifty years ago from Colette's provincial, earth-loving mother, the terrestrial Sido of "La Naissance du Jour," have been published by the literary weeklies for the first time, and also, like a third link in the female line of family sensibility, a paragraph from Colette's daughter, which terminates by quoting Saint Louis, the King: *"Merci, Mon Dieu, de m'avoir prêté Madame Ma Mère."*

February 17

It is too well-known a fact that governments in France fall too often. Everyone is also fully aware that there is a connection between government falls and the plethora of French political parties, of which right now there are eleven, six of them being major. To Americans, with their old-fashioned system of two parties, and even to the British, with their three parties, there seems to be something comic in France's having nearly a dozen. Obviously, the fact that there are eleven parties here is a tragedy. Since 1953 is bound to be filled with the acute argumentative decisions looming on both sides of the Atlantic, the prestige of France as West Europe's leader can ill support the strain of crumbling governments. Some sort of reform—under Premier René Mayer, if he doesn't fall before or during the attempt, or under whoever comes next—is regarded as inevitable, for to remain in power now is like trying to walk on marbles. The Fourth Republic's constitution, in memory of Marshal Pétain and in terror of General de Gaulle's position of solitary eminence just after the liberation, was expressly created weak, to prevent dictatorship. It was carefully written to give too much power to Parliament and not enough to the executive government. In Mayer's inaugural speech, he said he would single out for revision Article 13, which, if reformed, would permit a premier, under exceptional circumstances, to pass unpopular, needed laws (such as a law to tackle the French citizenry's tax evasions) by himself, thus relieving the deputies from having to answer for them to their angry electors; and Article 51, which, if reformed, would make it easier for a premier to dissolve Parliament. Then if a government fell, Parliament could fall with it, which, of course, would put the fear of God into deputies, and into political parties, too. But while these reforms may increase stability, they will not reduce the number of parties.

The Rassemblement du Peuple Français, General Charles de Gaulle's party, now has eighty-five deputies. In the 1951 elections, the R.P.F. became the largest Parliamentary party of France, by a margin of fourteen deputies. Recently it lost thirty-two members (and three affiliates), who bolted, or were ousted as heretics, after they had voted for Pinay's financial measures in defiance of de Gaulle's orders that his deputies sit like monuments, voting no to all government proposals. In 1946, General de Gaulle dramatically abandoned the Presidency of the Provisional Government in order to keep "the spiritual national investment," which he said the French Resistance and Liberation symbolized in him, from being sullied by party politics. The part of the French public and French politicians that has feared his later aims as being toward a *coup d'état* are still unable to explain why, if he desired dictatorial power, he did not seize it at the war's end, when France was at his feet. His R.P.F. political party was founded in 1947 not by him but by admirers, who gave it to him and the public like an extraordinary gift. What draws his followers is his conviction that France has rotted, socially, financially, and politically, and needs a drastic upheaval. Formerly, the General seemed to believe that a low-grade Parliament would bring about national collapse, which would thus allow what he calls *le pays réel* to vote him in as a savior. But since last month everything has changed. His party, originally "above party politics" and thus sitting in Parliament in a dignified vacuum, is now tentatively playing the parliamentary game. De Gaulle's aspiration is to be President of France, but a sort of idealized American President, with strong powers, a firm executive hand, and a Parliament reduced to a purely legislative role. As President, de Gaulle would treat the Communists, who obey Moscow, as treasonable criminals. He has consistently declared that the Fourth Republic's constitution demands a reform, which nearly everybody now admits. De Gaulle, as the incarnation of French patriotism, says that NATO has turned France into an American protectorate without adequate American protection; he wants France's defense run by Frenchmen, not by the United States. He is against a Federated Europe as being the death-blow to the sacred French nationality; what he does want is a Confederated Europe. General de Gaulle has lately been sending the Paris press frequent, lengthy communiqués on his newest political formulas. Even in addressing the newspapers, his language remains that of Bossuet, Bishop of Meaux nearly three hundred years ago. If the international situation deteriorates greatly, it is always possible that de Gaulle will make history again.

The French say that all their eleven parties and differing ideas—when they do differ—spring from the characteristic individualism of the French mind; that this double-edged genius for variety is the basis for the overdevelopment of the democratic principle of choice, now swelled through logic to the point of functional weakness. Even French deputies of the same party vote individualistically, as do some American congressmen. There is no party discipline, as in England, of obeying the whips. It is impossible to prophesy, for instance, how the French deputies and parties—except for the Communists, who will vote no, and the M.R.P.s, who will probably vote yes—will ballot when the European Army project comes up. French citizens who are disheartened by French politics think their politicians have become professionals, who deal in power as a business, like trading in eggs or real estate, as if their responsibilities were not the whole fabric of people's lives.

It is curious that the Third Republic, of which the Fourth is the descendant, started out in 1871 with hardly any party system at all.

March 10

Mournful singing by a Russian male choir preceded each of Friday morning's repeated Russian-language broadcasts, over the French National Radio, of Radio Moscow's announcement that had given people here the first news of Marshal Stalin's death. Then came Radio Moscow's French translation, accompanied by Tchaikovsky's "Pathétique" Symphony. Of the satellite capitals behind the Iron Curtain, Prague came in clearest over the air, playing the funeral march from Beethoven's Third. The French had listened to the same music with the announcement, from Germany, of Hitler's death and, from Washington, of Franklin Roosevelt's death. With anxious curiosity, the listeners wondered then, as they are wondering even more now, what would happen next in world affairs as a result.

L'Humanité—the only Communist daily paper left in Paris since *Ce Soir* was forced to cease publication last week, because of a dwindled circulation—perforce shouldered the Party burden of grief, propaganda, and official pronouncements attendant on the Marshal's demise. *L'Humanité's* Friday issue was a special edition of ten pages, with the front page bordered in black and a gigantic smiling photo of a Stalin of many years ago dominating the single column of text, which opened with "DREADFUL NEWS. STALIN IS DEAD. THE WORKERS

OF THE WHOLE WORLD, THE PEOPLE ALL OVER THE EARTH, ARE IN MOURNING." Thereafter, ironically enough, the Communist paper used, and credited, an American capitalist news agency, the United Press, for the death communiqué issued by the Soviet Union's Central Committee of the Communist Party, the Council of Ministers, and the Presidium of the Supreme Soviet—a communiqué that ended, as is customary when a European ruler dies, with cheers for the succession, in this case threefold and carefully impersonal: *"Vive notre puissante patrie socialiste! Vive notre héroïque peuple soviétique! Vive le grand Parti Communiste de l'Union Soviétique!"*

In its special edition ten pages along with a glorifying Stalin biography and an album of Stalin photographs, including the well-known happy one with Lenin, on a garden bench—*L'Humanité* reprinted, from the day before, as if insistent on getting the Soviet doctors' records quite straight, for once, the surprisingly detailed medical report on all Stalin's organs and the treatments used.

Samuel Beckett is an Irishman who came to Paris in the early nineteen-thirties in order to fall more completely under the sway of James Joyce, whom he already admired, and to whom he attached himself here. Since then Beckett has been best known for being a recluse who refused to meet people, and for his two novels, apparently written in French—"Molloy" and "Malone Meurt," which are in the Joycean "Finnegans Wake" manner and would be quite difficult to read even if they were in English. Now Beckett has written a curious and interesting two-act play that is being given at the Théâtre de Babylone, which is a typical little theatre hidden in a court behind an apartment house on the Boulevard Raspail. The play is called "En Attendant Godot." Godot is God, as nearly as anyone can make out. Those who are waiting for Godot are exactly two: a couple of tramps—one, named Wladimir, who is the spirit of wisdom, and the other, called Estragon, who is a weaker, coarser vessel. Across the heath where the tramps wait comes what looks like a decayed Irish squire, who may be the Devil or even might be Godot, whom the tramps don't recognize even when they see him. The squire drives and whips his old liveried servant like a horse. There is one minute when the servant, who may represent humanity, and who usually is dumb, suddenly starts talking at his master's order, spouting an insane hysterical muddle of erudition, formulas, and mental confusion that is one of the harshest, funniest satires on bogus learning since the satires of Swift. In Act II, the squire, who by now is blind, drives over the heath again but forgets he had met the tramps, and so drives off with his lunatic lackey, and the two tramps settle down again to wait for Godot, but far less sure he will ever come. Lucien Raimbourg, who plays the philosopher tramp, is a cabaret *chansonnier*. He is also a fine, simple, sensitive actor.

"En Attendant Godot" is written with an extraordinary sense of theatre, by which a drama with nearly no action and presented on an improverished small stage maintains acute interest in mere talk between people who do nothing but hope and get defrauded. Several of the critics have hailed it as one of the few truly intelligent plays in town. Certainly it combines many things. In its intelligence, it feels Russian, though done in the Irish manner and spoken in French.

May 6

If some of the French nowadays feel helplessly cut off from what goes on in the national government—in Parliament and inside their deputies' heads—most of the citizenry still feel close indeed to their local functionaries, the city or village mayor and councillors who do the neighborhood governing right under the citizens' noses. Accordingly, seventy-eight per cent of the voters actually voted the other day in the municipal elections in France's thirty-eight thousand communes. At least eleven councillors are elected in every municipality—even in villages with a population of a few hundred—which meant that all over France something like one voter in twenty-five was running for office. In small towns, mayors are paid merely a token salary, but they get to drape themselves in the official tricolor sash, which is worth a fortune in dignity; the councillors receive nothing at all. In a Hérault hamlet called Pégairolles-de-l'Escalette, all eleven councillors refused at the last minute to run again, because they had had too much trouble governing the one hundred and eighty inhabitants (and each other), so since Election Day the place has had no councillors at all, and heaven knows what may happen. At Montagnard-Dieu, in the Ardennes, all nine voters, who were using the mayor's cooking pot for the ballot urn, voted early so his wife could cook lunch on time. The daughter of Nancy's mayor was fined for nocturnally pasting posters around town that libelled her father's political rival. At Bessuéjouls, the mayor, who was up for reëlection, his wife, and one of his sons were all murdered during election week, and the addled local politicians failed to present a new candidate to take his place. More Balzacian incidents and characters are connected with France's municipal elections than with the voting for the national job of deputy, a big plum and sterner stuff. Among the special figures recently elected to office in their home towns were one member of the Académie Française; two actors, including one from the Comédie-Française; the president of the French Rugby Federation; American Ambassador C. Douglas Dillon's cousin Weller Seymour, of the village of Neaufles-St.-Martin; two Communists in prison, including Comrade Henri Martin of *"Libérez Henri Martin"* slogan fame; and the former noted Cabinet minister Georges Bonnet, who was not even a candidate.

The outstanding political result of the elections was the destructive defeat of General de Gaulle's party, Le Rassemblement du Peuple Français. Ever an unpredictable, Cartesian leader, in an amazing statement just printed in his party paper, *Le Rassemblement,* he dissolved the party itself. The old R.P.F. deputies and the few newly elected ones may no longer call themselves R.P.F.s. He does not disavow them as his *compagnons,* but as politicos they are now nameless, and on their own. (About thirty rebellious deputies of his achieved this personal liberty for themselves last year by backsliding out of the party.) The original salvational Gaullist movement, conceived to rescue France from the corruption of too many political parties, was itself transformed in 1947, by successful elections, into one more political party—against the General's wish. It is this party that, after these disastrous 1953 elections, he has just dissolved. The Gaullist Rassemblement movement itself he has by metamorphosis just restored to its exalted, unpractical form—a kind of patriotic political cult. Disdainful, as ever, of politics per se and interested only in the art of governing, the General scathingly said in his R.P.F. dissolution statement that in France "neither the Left nor the Right can govern alone. Acting together, they neutralize each other. The world sees this as it watches our sad political circus parade. The French regime is sterile and for the moment cannot change—a series of combines, bargains, majority votes, and office investitures that make up the games, the poisons, and the joys of the system." But he foresees, as he has been foreseeing for the past seven years, that France is facing a serious shakeup, and that with it will come what he bitterly called "the bankruptcy illusions." "The safety of France," he continued, "and the state will be at stake. *Il faut préparer le recours."* And who will furnish the means that will save? Le Rassemblement, whose ex-R.P.F. candidates just lost the municipal elections.

"Le Salaire de la Peur" ("The Wages of Fear"), the French film that won the Grand Prix at the Cannes Film Festival, is now being hailed by Parisians as a truly great French film. It is certainly a great cinematic thriller. There are three hours of it, focussing on Georges Arnaud's brutal best-seller novel of the same name about international riffraff rotting, unemployed, in some Central American oil region until heroism becomes a paying proposition. A distant oil well has caught fire, and explosives are needed to blast out the flames. A tough American oil agent offers a fortune in cash to the four bravest barflies to drive two trucks of nitroglycerin over a dangerous mountain road to the conflagration.

The too long first half of the film is like a remarkably photographed novel, showing the barflies' sordid, greedy quarrels as a guide to what is coming. From there on the shocks begin, with a fearless, fat Italian and a disillusioned, heroic German in one truck and two Frenchmen, a lovable coward and his murderous, gay young friend, in the other. The mountain ride, which is what the film was made for, becomes the acme of anguish, a kind of motorist's nightmare, with a new twist of sadistic slowness, since the trucks dare not go fast. The decaying bridges, the precipices, a landslide, a near drowning in an oil pool, the deaths, and even the temporary victory are exploited in tension, inch by inch. Only the murdering young Frenchman arrives alive to claim his prize money. To the music of a remembered juke-box waltz, he then drunkenly waltzes his empty truck into a chasm on his return trip over the mountain road. The wages of fear have been paid.

The acting is excellent and the international principals—as well as the director, Henri-Georges Clouzot—deserve mention by name: Folco Lulli, as the Italian; handsome Peter Van Eyck, as the German; the late William Tubbs, as the American oil agent; Charles Vanel, as the white-feathered Frenchman, for which performance he won a Cannes acting award; and Yves Montand, ordinarily France's most popular tough-guy music-hall ballad singer, as the gay murderer. "Le Salaire de la Peur" is a superior character film with anguishing thrills.

* * *

The European antics of Senator McCarthy's youthful investigation team, Roy Cohn and David Schine, gave Parisians rare material to sharpen their wits on. Weary of hearing about American efficiency (and tired, too, of trying to find some comforting evidences of it in certain of the slapdash postwar American installations flourishing around France), the French went into stitches of laughter over a report that Cohn and Schine investigated the loyalty of twelve hundred employees of Radio Free Europe in Munich in thirty-five minutes. They thought it the greatest half hour of mass inquisition ever known in Europe's two thousand years of political practice. According to American diplomatic gossip from all over the Continent, the Cohn-Schine Katzenjammer Kids, which is what Americans here called them, have done more and funnier harm to the prestige of the United States in Europe than anything else that has come out of the astonishing McCarthy box of surprises.

What started three weeks ago as a small strike in the upholstery department of the nationalized Renault motorcar works in the suburb of Billancourt spread into a strike of ten thousand workers that has turned into a lockout with thirty-seven thousand men idle. The strikers demand twenty-five francs more an hour. There have been strikes in several of the big hotels—the Crillon, Meurice, Lotti, and Continental. The Continental's strikers marched over to the sunny Tuileries Gardens the other morning—the chefs in their tall caps, the waiters in their tailcoats, the bartenders in their white coats, the dishwashers in their dungarees, the *garçons d'étage* in their colored jackets, and the chambermaids in their billowing aprons—after kindly serving early-breakfast coffee to the guests, who thereupon proceeded to make their own beds. There is talk of strikes in the gas and electrical companies and in the railways, to tie up the coming businessmen's Foire de Paris. It is hardly a secret that most French labor is underpaid. Labor has apparently decided to let French business know, too.

May 20

For the past month, American visitors to Europe who are important enough to know important Europeans have been reporting that they are everywhere being asked the same anxious, two-headed question: "What are you going to do about McCarthy? What has happened to America?" In the past week in Paris, Senator McCarthy has suddenly become the main Franco-American topic of shrill conversation. The Paris newspapers, which at first were worried about McCarthyism, are now profoundly afraid of it and have been writing about it angrily. A recent typically complicated French headline was "MCCARTHYISM, FORERUNNER OF FASCISM IN THE USA; SENATOR IS LEADER OF PRO-FASCIST GROUP WHOSE CENTER IS IN CHICAGO (NEAR WISCONSIN)." A typical Paris editorial said, "In his indecent attacks on those he dislikes, insult is mixed with excommunication. Goebbels in his best days did not do it any better." *Franc-Tireur,* which has always been pro-Eisenhower, said in an apprehensive headline, "MCCARTHY IS COMPROMISING EISENHOWER'S FUTURE," and referred to the Senator as a scarecrow with a bloody mailed fist. The conservative *Monde,* the most influential newspaper in France, had as a headline "THE MANIAC OF THE WITCH HUNT; DAILY HE WEIGHS MORE HEAVILY ON THE LIFE OF THE AMERICAN PEOPLE."

Le Monde has given the most space, voice, and variety to the French concern over McCarthy. In an article called "The Two Troikas," it mentioned Moscow's three-horse government of Malenkov, Beria, and Molotov and asked, "Who reigns?" Then it took up the Washington triad of Eisenhower, Taft, and McCarthy, ending with the query, "Who governs?" *Le Monde* added, "The President of the United States should not forget that he is the leader of the free world, a title to which neither Taft nor McCarthy can pretend." And in an analysis of what it termed the danger to freedom of thought, which is what most alarms the individualistic French, the paper said, in part, "The drama of American liberals in their opposition to McCarthyism is that their counterattack aims at his methods rather than at the underlying issue in McCarthy's investigations. Nobody dares to raise the true question in the political and juridical fields: Can the Congressional investigations establish control over the political and religious ideas of private American individuals?" Of all the Paris press this last fortnight, the Communist *Humanité* alone has not lifted one stick of type against Senator McCarthy.

June 1

This is the time of year when French country cousins come up to Paris from the provinces and their estates to attend the June-season horse races, balls, family dinners, and match-making tea parties, and to take in the major painting exhibitions as their annual snack of art. The favorite exhibition right now is a luscious one of nudes at society's preferred gallery, La Galerie Charpentier. It will last all summer and is obviously an ideal hot-weather show. So much nudity on the walls presents an astonishing sight, with appetizing, rosy torsos and limbs from the days of the Ecole de Fontainebleau, a classic nude by Poussin, a goddess nude by Chassériau, and modern nudes by Renoir and Cézanne. It is a stimulating show, if only because it reveals how the human form has been bred during a four-hundred-year studio evolution, in which pretty, pink, round women have become thin, slab-sided, or even blue-breasted ones, with angles.

People here who have been worrying more than usual about the social unrest rising in France have just received an unusual, dispassinate analysis of it from an unexpected source. Monsignor Paul Richaud, Archbishop of Bordeaux, last year headed an inquiry into the possible causes, undertaken by the bishops of Catholic France. *La Voix Diocésaine de Besançon,* a religious review, recently gave an outline of the resulting report, which has provoked considerable amazement and comment on all sides. After the bishops had conducted a detailed examination of working and living conditions for the proletariat in all the dioceses of France, they concluded that though the bourgeoisie and the owner classes feel a certain desire to uphold human liberties, the dominant mainspring of their actions is the profit motive. According to the report, these classes show an unconsciousness of the disproportion between the way they live and the way their employees live, and they do not suspect that their comfort and riches prevent their lucid appreciation of the workers' problem. The idea of fair prices is losing ground, profit margins swell, and the sharing of profits with workers is almost nonexistent. As for the workers, the bishops went on, it is a fact that they always have to use force to improve their lot. Most are convinced that they are victims of organized injustice. Their salaries, even with social insurance and family allocations from the state, are proportionately less than before the war, though the national income has gone up. Workers, the bishops found, in many cases think that they have some rights in private business that today's society will not recognize. Though most of them—or so the churchmen thought—do not accept the Communist theory of class struggle, they recognize class struggle as a social fact, and their acute comprehension of the human rights denied them by the owner and bourgeois classes is back of their impatient desire for reforms in the social structure. For, the bishops concisely concluded, the French worker is more hopeful of changing structures than of changing the habits of the Frenchmen above him. French commentators are as impressed by the candor of the Archbishop's report as by the Church's obvious concern with France's present social malaise. Plenty of French also think that if the Church aims to use its vast influence for social betterment, it cannot start using it too soon.

The eighteenth French government since the Liberation having fallen last month, it was thought by many here that the political program demanded by the third of the premier-designates who failed to form a nineteenth government—little Paul Reynaud—was essential. He said that it was nqt worth while taking office unless the constitution was first reformed to assure a French government of eighteen months of governing, instead of the postwar average of five months, and that Parliament should bear the punishment and responsibility of dissolution. Until then, he said, France cannot be consecutively governed. Until then, as the French citizens helplessly see the tableau, each French government is merely another balloon that barely gets off the ground before it is snagged on the Parliament roof.

June 18

France has been four weeks without a government, although nine politicians have been offered a chance at the premiership by President Vincent Auriol. An acid newspaper cartoon by blasé old Sennep shows Auriol wondering whom to ask next and murmuring, "Maybe a well-known woman—Danielle Darrieux or Martine Carol . . ." Assembly President Edouard Herriot, aged eighty-one and sick in bed, received at his bedside his Radical Party leaders, whom he belatedly warned to stop playing party politics, saying, "At the close of my life, with all my soul I implore you to think now of nothing but France." The politicians had better think of their own skins, too. According to Article 51 of the Constitution, this Parliament can be dissolved and the deputies sent hunting for their jobs in general elections if they overthrow one more government on a question of confidence or censure between now and November, 1954. To the French public, the present crisis fundamentally shows just what the seventeen other crises since the Liberation have shown, but with a strange air of finality. It shows that if a French Premier is to govern with any continuity today, he must be given special powers. For the first time in a crisis, there is practically a sitdown strike among designate-Premiers, who will not lift a finger unless they are promised better Premier working conditions. Above all, this crisis, like a magnifying glass grown seventeen times stronger, makes ultra-visible the fact that France's multi-party system and her disparate ways of looking at political matters—a disparity upheld by French logic and French republican liberty—have reached a point where, through habit and through shortsighted human psychology, political greed, and Gallic precedent, party minded men, both in and out of politics, are concerned only with their differences in thinking, instead of with developing the ideas they have in common.

Considering all the pressures besieging France—the delayed Bermuda conference, the old unsettled business of the European Army, the new influential, softening Soviet policy that almost daily is seeping out from behind the Iron Curtain, and France's vital need for a revived diplomacy and some set foreign policy, even about its own war in Indo-China—considering all these exterior semi-crises, the ridiculous aspect of this present crisis is that it began in France's hip pocket over a question of money, although France is the richest country in Europe. There has always been the comforting theory that France's financial troubles come merely from the fact that her citizens don't pay their taxes. According to the last Finance Minister's Report on the Economic Budget of the Nation, the state revenue would admittedly be billions of francs higher if the citizens paid all the taxes they owe—that is, if French businessmen did not classically keep two sets of books and pay only on the lesser book. But the report says, nevertheless, that more than a third of the national income does wind up in the tax channels. Yet with all that money rolling in, Premier René Mayer had to borrow money from the Banque de France, which, in a way, is why he fell and why the crisis began. Then the Banque de France, dunning for a payment on the debt, knocked on the government's door and found nobody home.

June 30

If—as was certainly not the case—France had been able to afford her recent record-breaking five-week political crisis, during which she was without a government, the crisis would have done almost nothing but good because it so harmed the present kind of parliamentary regime in the estimation of the French people. Unbearably costly as the crisis was in lost prestige and even financially, it seems, at any rate, to have proved to France's citizens a fact they had to make sure of, once and for all—that the present constitution absolutely must be reformed, so Parliament will not have the power to do again what it has just been doing, and so the executive will be able to sit in the seat of state long enough to govern. According to experts, if anything is sure right now in French politics, after the incomprehensible choice of an obscurity like the Calvados deputy Joseph Laniel as Premier, it is that Parliament will undertake a revision of the constitution, if only to save itself from possible dissolution.

The outstanding salutary human element to emerge during the critical month was the bold, harsh intelligence and the uncompromising personality of the youngish politico Pierre Mendès-France, as revealed in his unaccepted program for saving the country through economic and social reforms. Plenty of French have got into the habit of thinking that their country cannot be saved. Mendès-France's program, outlined in the speech he delivered in Parliament when he was trying to get support for his bid for the Premiership, set a standard by which all other programs have since been judged. Though the parliamentary majority turned his program down, millions of French accepted it in their own minds and remember it as the one notable declaration among the millions of words of *blah-blah-blah* (French for "yackety-yack") spoken here. A lawyer and a banker trained by the Rothschilds, yet a representative of the younger-generation reform wing in the liberal, if money-conscious, old Radical Socialist Party, Mendès-France remains the dominant figure of the crisis and, with luck, he could be the new leading man of France, which certainly still needs one.

The crisis reminded one of Alice and the Red Queen when they started running, not to get anywhere but merely to stay on the same spot. With the crisis ended, France is just where she was before it began, though more winded. As ex-Premier Antoine Pinay, one of the unsuccessful candidates, remarked bitterly, there are two hundred and five Marxists on one side of the Assembly, and, on the other, four hundred and twenty-two anti-Marxist deputies, whom it is impossible to pull together into an unselfish, non-party national coalition that could put France back on her feet. And as the semi-Socialist afternoon paper *France-Soir* pointed out, still to be solved are France's interior and exterior financial deficits, the European Army problem, the Indo-China war, the Moroccan troubles, rising unemployment, inadequate salaries, a stagnant economy, investment difficulties, and insufficient housing. The height of Parisian editorial indignation at Parliament's feeble choice of the inexperienced new Premier—after its five weeks of lionlike roaring—was reached by the independent *Combat.* "The cowardly relief felt by the deputies in seating M. Laniel is not shared by the rest of the country," it fulminated. "They are making another grave mistake if they think it is."

During the five critical weeks, the Paris newspapers kept printing, if merely from habit, what the parliamentarians thought about politics. Only one paper extensively printed what the French citizens thought about Parliament. This was *L'Express,* the new Paris political weekly edited by Jean-Jacques Servan-Schreiber, who previously was one of the noted younger political experts writing for *Le Monde.* What he printed in *L'Express* was a culling from the famous regular reports on what the French feel and think about life and politics which are gathered (according to a shrewd system started by Napoleon Bonaparte) by the prefects' offices in each of the ninety departments of France; the reports are sent to Paris and boiled down into a synthesis, which is available to three officials—the President of France, the Premier, if there is one, and the Minister of the Interior. *L'Express's* own editorial analysis of its selection of nationwide prefectural reports was that the French are so sick of Parliament they might even accept some single leader to save France "if only he gives an impression of being honest, brutal, and courageous." Unquestionably most of the citizens' opinions, as represented in the prefects' reports, were scathing. In and around Rouen, people wrote three times the average budget of monthly letters to their prefect, all full of complaints. They charged that their deputies had shilly-shallied deceitfully in abstaining from voting on all the candidates for Premier put before Parliament, and in not stating clearly on the Assembly floor whether they were for or against Mendès-France's program, and why. They also said that Parliament ought to be dissolved, but only after passing an electoral reform, since "otherwise the same deputies would turn up again and the muddles would continue." In Compiègne, a leading citizen declared that "the deeply disgusted French people demand a dictator with a pitchfork and broom." In Boulogne, a businessmen's petition asked that "the nation choose new representatives, who would really work to restore France." In Lyon, home of Parliament's president, old Edouard Herriot, there was an increased wave of antiparliamentarianism. In the northern coal region, workers thought the high cost of living more important than politicians, "who never change anyhow." In the south, the major theme was again dissolution of Parliament, preceded by electoral reforms to prevent the reëlection of *"les incapables,"* along with prophecies of serious social troubles this autumn if there is not a new government atmosphere and crew. The prize reaction to the crisis came from a citizen of Marseille, who wrote to President Auriol (the President declared that his bulging private letter bag showed him how dangerous and profound is the malaise of the French people) demanding the dissolution of Parliament and of all eleven political parties in favor of a tricolor format of only three authorized groups: a blue party for the Republicans, or right; a white party for the center; and a red party for all the left. It was a sensible, if impossible, plan.

The major post-crisis worry now is just plain money. According to the caretaker government's report, the crisis cost France the equivalent of two hundred and eighty million dollars in unfloated loans and other losses. The state coffer is bare and is taking another massive loan from the Banque de France to pay state employees their end-of-June salaries. What the French people now want to know is by which of the two customary agonies will the deficit be met—by extra taxation or by printing more inflationary money.

Obviously, France has had so much history that it necessarily keeps repeating itself. Apropos of today's terrible deficit, there has been recalled, like an echo from the past, Louis XVI's financial dilemma, during which Marie Antoinette supposedly asked the Royal Finance Minister, "What will you do about the deficit, Monsieur le Ministre?" "Nothing, Madame," he replied. "It is too serious."

* * *

The Soviet violinist David Oïstrakh, officially described by his government as Russia's greatest, was enthusiastically accepted here, after his concert at the Palais de Chaillot, as one of the great performers from any land. A stranger to Paris musicians and an inaugural export in the new relaxed Malenkov policy toward the West, he packed the theatre, the largest in town, with political comrades, with music lovers, and with curiosity seekers. Oïstrakh is a nice-looking, big, blond, stocky Ukrainian with an undershot lower lip and extremely long, lithe fingers, which, as regards his left hand, may explain his impeccably accurate tone. He keeps his bow arm close to his side, produces a *gamme* of sounds that are beautiful—especially the soft ones, including unscratched pianissimi in alt—and displays no folderol mannerisms. Of the three concertos he played, the opening one, Mozart's No. 5, was unfortunately led like a cart horse by Jacques Thibaud, exceptionally directing the Conservatoire Orchestra, which was afterward vigorously managed by Manuel Rosenthal. The final work, a Khatchaturian concerto dedicated to the player, was a good-natured, folksy affair, with an amorous *valse lente* as its *andante sostenuto,* which, by giving merely entertainment, could offer no notion of Oïstrakh's important qualities. It was in between the two, in Brahms' Opus 77, that he presented his rich, complete musical report. Controlled and unified, his emotional interpretation seemed universal as well as particular; his musicianship serious, learned, and positive; and his technique a matter of almost luxurious ease, but used unobtrusively and without the counterfeit of tricks. At the concert's end, he received an ovation, the audience clapping without pause for a quarter hour, until he delicately played, with more courtesy to France than to musical taste, a violin arrangement of Debussy's "Clair de Lune." According to the program notes, Oïstrakh, who wore his Stalin Prize medal pinned to his evening clothes, was born in 1908 in Odessa, the son of an opera-chorus singer, and was a pupil of Pyotr Stoliarsky, who seems to have trained all the past generation of Soviet violinists. Since 1934, Oïstrakh has taught at the Moscow Conservatory. At his concert, he played a Stradivarius bestowed on him by the Soviet government.

August 18

Symbolic and descriptive of Paris on the fourteenth day of France's paralyzing national strikes was the long line of municipal refuse trucks driven by French soldiers and rattling past the empty, closed Parliament buildings—the Army en route to its matinal job of collecting city garbage while the deputies were off somewhere on Parliamentary vacation during the greatest civil-service crisis of the Fourth Republic. This is the biggest strike movement France has experienced since the violent Front Populaire days of 1936, and the only calm one of consequence in French history. So far, not one of the many sides involved—strikers, police, labor unions, political factions, the military, and the discommoded public—has lifted the theoretical or physical finger necessary to cause or incite bloody riots. No one—or nearly no one—on the Paris newspapers has raised his voice in the usual shrill, tantalizing editorial. As one ordinarily tendentious commentator wrote solemnly, "This is an enormous, majestic strike."

It began in the smallest, most careless way possible, according to two remarkable articles just printed in a couple of rival Paris weeklies. When the stories are put together, what they say is that the strike started because of two men and a typing omission of four words. These men were MM. Landy and Fossat, employed in the money-order section of the Bordeaux Central Post Office. At four-thirty on the afternoon of Tuesday, August 4th, there mysteriously came into their possession a mimeographed advance copy of one of Premier Laniel's economy projects, compiled by Edgar Faure, the Minister of Finance and Economic Affairs, which, among other things, listed the categories of civil-service employees whose retirement would be postponed for five years. Because four words, *"facteurs et facteurs-chefs* [postmen and chief postmen]," had been omitted, through a typist's error, from the stencil, postal employees were not on the list of *actifs;* that is, workers who, "because of fatigue in performance of service," were to be spared the delay in retirement. The postal workers thus appeared to be transferred to the list of *sédentaires.* This, to Landy and Fossat, seemed no place for foot-weary mailmen. At five o'clock, after consulting the local higher-ups in their union, these two men called the entire Bordeaux P.T.T. (Postes Télégraphes et Téléphones) service out on strike. At eleven the next morning, after the pair had talked by phone, which by now nobody else in Bordeaux could do, to the top leaders of the union in Paris, the national postal strike was called. When the Paris union leaders later sought the customary interview with the appropriate government Minister, they discovered that the Minister of the

P.T.T., Pierre Ferri, was in Lisbon to report to an International Postal Congress on France's fine postal service, and that the Minister of Reconstruction and Housing, Maurice Lemaire, who was assuming his functions in the interim, was completely ignorant of the Faure project's specifications. He also did not know how on earth Landy and Fossat could have come by a document that bore the stamp of the Minister of Finance and Economic Affairs. Investigation traced it to the Conseil Supérieure de la Fonction Publique, whose director, by law, must submit proposed decrees to the council members, twelve of whom, also by law, are labor-union representatives. One of these must have passed along the copy that somehow ended up in the Bordeaux Post Office. As soon as the project's original text had been checked at the Fonction Publique office, the omission of the four fatal words came to light, but it was too late. Twenty-four hours after the postmen struck, about two million other government employees began joining them in sympathy strikes, which by midnight on Thursday included transportation everywhere, and France was atrophied.

In Paris, the paralyzing effect of the strikes was increased by the lethargy of La Fête de l'Assomption, the big religious and bank holiday, which seemed bigger than ever because it fell on a Saturday this year, and on Monday the newspapers were also on strike. Boulevards are empty, hotels are stagnant, many restaurants are locked for their annual August closing, but enough are open for the reduced clientele, with fresh food brought in to town by country camions. The gas strike has merely made pressure low. As for electricity, Paris, being the Ville Lumière, still has it to burn, with the neon advertising blazing nightly, and this last weekend Notre-Dame and other sites were treated to nocturnal illumination. The Louvre, like all museums, is closed, the guards being civil servants, and thus on strike. The streets are cleaner than New York's usually are, because of the presence of soldiers equipped with trucks and brooms. On the Métro, now running more frequently, though still stopping at 6 P.M., we often ride several times on the same ticket, because ticket punchers are on strike. The trains don't stop at all stations, or even at all the stations they are supposed to stop at. You get out where you can and walk back, or go to the end of the line and hope for better luck at making your particular station on the return trip. Everybody walks a lot. The Army has saved many people's legs by substituting Army trucks for the striking city buses, but they stop at nine at night. An operational center has been set up on the Place de la Concorde, near the Strasbourg statue, with a map of Les Lignes Intra-Muros Exploitées par l'Armée leaning against a lamppost, and a two-way-radio truck run by an efficient, hard-boiled corporal of the Compagnie Républicaine de Sécurité, toughest of the French police shock troops. The Army trucks have a couple of boards, set lengthwise, to sit on and usually no top, so standees can be packed in more easily; are driven by a soldier; and have a soldier on the tail to hoist passengers on and off. Bus tickets are bought by the book at tobacco shops, for the soldiers are not supposed to handle cash. In front of the Grand Palais, a motley collection of privately owned buses, ordinarily used for tours, weddings, or funerals, are now operating to carry people at least in the cardinal directions away from Paris. Inside the Palais, the ticket sellers are flanked by blackboards with *"Nord,"* *"Est," "Sud-Ouest"* written on them in chalk, as mere points of the compass for the travellers to head toward, with the specific town, like Caen or Cahors, that the bus is going to added almost as an afterthought. Waiting passengers—each day they are fewer, but they shrewdly come hours early to be on the safe side—sit on the Palais steps, tidily piling their bread crusts or chicken bones in the crevices of the façade sculpture. Mothers sprawl on the grass beneath the trees, nursing or changing their infants. Aside from the ticket sellers, the huge interior of the Palais is empty while the patient citizens sit outside in the sun or occasional rain.

The Palais bus schedules, and the street addresses and bus schedules of half a dozen private tourist agencies that are also providing transportation, are printed in the newspapers and announced during the news broadcasts of the national radio, which is operating again. The radio also announces personal tragic news, such as that the mother of Mme. Telle-et-Telle, of such-and-such a place, is dying, and agreeable collective news, such as the good health of children in the hundreds of vacation colonies all over France.

September 2

The strikes are over, nearly everybody is back at work, and nothing is settled. It is now figured that at the peak of the strikes four million French—or close to a tenth of the entire population, including infants in arms—were out. Yet the French themselves say that after this vast social upheaval it is not clear who won—or, indeed, whether anyone won. Only the losses are easy to see. The government lost twenty-five billion francs in revenue from its paralyzed railroads and bureau of Postes, Télégraphes, et Téléphones, which is reportedly just about what it had hoped to save on the economy projects that set the strikes going. Masses of workers lost three weeks' pay while they were striking against not getting enough pay in the first place. Now the government says it will allow certain categories of workers, such as the postal employees, who started it all, to make up part of their lost pay in overtime, on wartime work, which has collected all across the nation like unwashed dishes. Since most of us are now getting a dribble of letters mailed a month ago, there seems to be plenty of overtime cleaning up for the postmen to do.

Though *Figaro,* as the upper-class morning paper of Paris, was cautious in its editorials during the troubles, it has now printed an exceptionally frank warning to the government, and to private enterprise, too, which says, in part, "Let us not forget that the explosion of discontent was provoked principally because numberless workers once more arrived at the conviction that they would get no satisfaction unless social agitation brought about such a state of urgency that certain problems would have to be looked in the face, if not solved. What is the use of denouncing the very real harm the strikes did to the country if strikes still seem to workers to be the only way of getting anything done and if we wait to do anything until the strikes break?" *Le Monde,* the upper-class evening paper, has just published a statement made by the Action Catholique Ouvrière, the remarkable organization of worker-priests, which says, "For years the French working class's legitimate hopes (and they are not solely material) have been betrayed. Contracts to which it was a party have been torn up or ignored. Workers' living conditions are becoming increasingly precarious, while a small number of the rich and the pleasure-loving make a show of insolent luxury. It becomes less and less debatable that the present French social, economic, and political structures are largely responsible for social injustice, for an inhuman economic system, and for class politics that are contrary to the very spirit of the Gospel . . . and the city of justice and brotherly love." Very unusual words to read in the City of Paris today.

* * *

A flock of important events occurred during the strikes, when the scattered or immobilized Parisian public had neither sufficient concentration nor sufficiently ample news to take in the strangeness of some of the things that were happening. Among these was the unexpected absence of bloodshed in Morocco when the old pro-Moroccan Sultan, Sidi Mohammed ben Youssef, was deposed and virtually kidnapped, and the absence of shock at the installation of the new pro-French ruler, Sidi Mohammed ben Moulay Arafa. With the unpopular, insoluble Indo-Chinese war, with the recent one-day visit of Vietnam's Emperor Bao Dai to President Vincent Auriol in his summer quarters at the Château de Rambouillet, with the present sharp Cambodian independence demands, and now with the switch of sultans in Rabat, France's colonial troubles have certainly reached their apex. Over the last few years, much of liberal French opinion considers that the colonial tableau has been altered, willy-nilly, for all European nations, even France, by the change in England's former empire. An outstanding Parisian leader of this critical group is the novelist and Academician François Mauriac, who sees his connection with the Moroccan problem largely as an extension of his Christian conscience. He has boldly stood out against the government's colonial policies, and has been an ardent backer of the native reform movement. Mauriac bluntly termed the recent investiture of the new tame Sultan a plot in which, in his opinion, the present French government was one of the conspirators—strong stuff indeed. In the August number of *La Table Ronde,* the monthly magazine Mauriac edits and of which his column called "Bloc-Notes"—excerpts from his personal diary—is always a special feature, there is a painfully timely illustration of the violent bitterness a certain ill-educated, hard-bitten type of French colonial feels against any liberal French reform views. The August "Bloc-Notes" mentions threatening letters Mauriac has lately received—in particular, a brutal one from Casablanca, which the magazine quotes, including words unprintable in English. "You might as well know," the letter says, in a quotable part, "that if you continue to write such tripe [favoring the natives] you are going to have an accident one of these days. You're a dirty dog and a traitor to Marshal Pétain. Take care, because you are going to be beaten up. My fists are in fine shape, those of a colonial farmer who spews on you all his contempt. You call yourself a Catholic and Christian. Talent isn't everything. There is soul, too, and you haven't any, and your spinal column is as supple as a cobra's. In a few days, it's going to hurt you good. *A bientôt, salaud.*" It is on such conflicting points of view among the French that the new ben Moulay Arafa administration in Morocco has been founded—the most important French event that has taken place outside France this year.

October 1

The autumn symphony-concert season has opened with Beethoven's Ninth and the "Eroica," and with Debussy and Ravel as usual. Some more freshly and vigorously appreciated concerts of jazz music have also just been given. American jazz—when Parisians can obtain the real article—is still of concert-size significance here. Since the 1948 concerts in Paris of Dizzy Gillespie, fans of *le jazz-hot* have starved for the sound of a big-name jazz band. Now they are almost having a glut, with five jazz concerts in the last ten days, and with more, by Count Basie, soon to come. Stan Kenton and his band gave one sellout performance at the Alhambra, and Lionel Hampton's orchestra gave four concerts in two days at the Palais de Chaillot, the largest theatre in Paris, which he packed. French *jazz-hot* fans are as intellectual in talking of *swingue* as if they were discussing Schoenberg. They praised Kenton for—as one fan put it—"his meticulous orchestrations, the maximal efficacy of his howling trombones, and his whistling cornets, which bathed you, drowned you, stiffened you with sound, till they suddenly gave way to the soft languor of saxophones, which were like balm." Kenton was also thought to be too intelligent in his jazz, too rehearsed, and too ascetic.

There was no intellectualization at all at the Hampton concerts, especially after the intermissions, when some of the musicians came down into the aisles to play and some of the audience climbed up on the backs of their seats to cheer. Hampton was, however, critized as a formalist for his fidelity to elements of boogie-woogie. It was his tenor-saxophone soloists, his quintet made up of a flute, three trumpets, and a sax, his vibraphone solos punctuated by his odd, bleating voice, and, above all, a few melodic climaxes of wild octaves run up by the brass in shrill arpeggio that brought down the house. The costumes of the musicians were also much admired. They wore violet tuxedos, huge Byronic shirt collars, and bullfighters' string ties.

October 14

In 1927, the French poet and diplomat Paul Claudel wrote a spectacle called "Le Livre de Christophe Colomb" for Max Reinhardt, of Berlin. Knowing what the Germans like, Reinhardt stipulated that it should have orchestral music, a choir, a ballet, and two Christopher Columbuses—one the old, friendless navigator on his deathbed, the other the young discoverer in his prime, with the new, round world and glory just in sight. For various reasons, the opus has had to wait all these years for a production, now courageously undertaken by the theatrical company of Madeleine Renaud and Jean-Louis Barrault, who opened in it upon their recent return to the Théâtre de Marigny, after a year and a half of foreign triumphs. The only simple element of "Le Livre de Christophe Colomb" is the stage set—a vast sail, which sometimes represents a ship, at other times, when semi-furled, a white curtain at Queen Isabella's palace, and later a cloud in Heaven. Otherwise, the play is utterly complex. There is a prologist, in modern evening clothes, who reads history from Columbus's old notebook, and there is a supporting cast of thirty-two, who cluster on stairs leading to the orchestra pit, where they sing Darius Milhaud's exhilarating music or shout, as enunciators of public opinion, either for or against Columbus. Occasionally, after donning cloaks or mustachios, they swarm up onto the stage as Spanish courtiers or sailors. There is a ballet at the Spanish court, with dancers representing Envy, Ignorance, and Avarice, and later a ballet of natives of the Azores. Much is made of a pious pun on the name Colomb, which is supposed to mean *"colombe"*—French for "dove"—and thus identify Columbus with the Holy Spirit. Aside from a storm at sea and a sailors' mutiny, little dramatic action is furnished by the lines, which are full of atmospheric repetitions that approximate poetry. By means of brilliant stage direction, Barrault has supplied at least theatrical action, using his huge cast to catch the spectator's eye when the spoken lines fail to attract his ear. The play ends with Queen Isabella, the two Columbuses, and a pair of saints being hailed with hallelujahs in Paradise.

"Le Livre de Christophe Colomb" has been positively eulogized by the critics, only one of whom found anything to regret. He said that he thought some members of the audience didn't like it. Lots of members of the audience didn't.

October 27

A week ago, the French Parliament, for once, actually voted almost unanimously on something—to hold a full-dress, decisive debate on France's vitally pressing, two-headed Indo-Chinese problem. When the debate opened last Friday, two thirds of the deputies failed to turn up, either to talk or to listen. Only the Communist members attended in full fig. The other parties' absentees were visiting their constituents to ask what they wanted them to say when the debate reaches its climax, as it will very soon. Certainly the answers to be given are extremely important not only to the French but to the rest of us. The two main questions are: What should France do in Indo-China about its French Union (formerly called the French Empire), which the Vietnam Congress in Saigon has just said it no longer wants to belong to—in its present form, it added hastily and too late. And what can or should France do about ending its unpopular Vietnam war? Already the Parliamentary group around Pierre Mendès-France, who is considered the new white hope of political realism, has asked that negotiations be opened with the Communist Vietminh leader, Ho Chi Minh. Expert civilian interpretation of the Army communiqués on the recent, much touted French-Vietnamese offensive south of Hanoi says that the military situation is actually no better than the political one, which is double poor news. Furthermore, from one end of France to the other the public is still extremely angered by what it considers the ingratitude of the Saigon Congress's recent demand for national independence (which, after all, France last promised in its note of July 3rd). The way the French see it, for seven years the cream of their country's manhood—its young volunteers and its Army officers, who are being killed at almost the same rate that new officers are being commissioned from St. Cyr—have been fighting and dying in Indo-China. Not only the popular indignation against the Vietnam Congress but also the possible opportunity for France to wash its hands of Vietnam in consequence were skillfully keynoted by the moderate-Left Paris newspaper *La Libération* when it said, "Those Vietnamese we have supposedly been fighting for now want the French to leave. In these conditions, what the devil are our soldiers doing out there? Why not bring them home immediately?" *Le Canard Enchaîné,* the satiric weekly, ran a cartoon that showed Marianne standing before a wall in Indo-China on which "Go Home" was scrawled, the way the French scrawled it on walls here for the Americans.

Inevitably, there has been more than the usual anti-Americanism in the press, provoked by the Washington administration's wish for France to continue bleeding against Asiatic Communism into the eighth year while we hold parleys for our political truce at Panmunjom. But some sensible observations have also been made. One concerned the common difficulty the French and the Americans, with their Western brains, have in dealing with the Eastern mentality. Both Emperor Bao Dai and his High Commissioner, Prince Buu Loc, have been right here in France during these recent important days, but for all the French officials could do with them, they might as well have been eating rice back home in Saigon. As ex-Premier René Mayer cynically remarked, "Probably the Vietnam Congress is no more slippery for the French to deal with than Syngman Rhee is for the Americans."

One of the few deputies who turned up for the Indo-China debate informed his handful of listeners that "the campaign of Maximilian in Mexico, which was one cause of the ruin of Napoleon III's army and the French defeat by the Germans in 1871, was a chef-d'œuvre of diplomacy and strategy compared to the absurd Indo-Chinese campaign, which swallows up all our Army cadres, prevents any straightening out of our domestic financial affairs, and annihilates our European defense program." Once more, the watchful French citizens are waiting to see how their Parliament, after its preliminary blowhard speeches, finally votes on the complex Indo-Chinese question.

November 10

One hears little any more of Les Six, except for the four who, since their chance union in 1918, have continued to write the major part of France's modern music. Les Six, if taken together, are Louis Durey, whose name is always frankly forgotten; Mme. Germaine Tailleferre, whose name is remembered; and Auric,

Poulenc, Milhaud, and Honegger, whose names and works are current and constant. Reassembled by Jean Cocteau, who made an almost paternal speech from among the Conservatoire orchestra's first violins on the stage of the Théâtre des Champs-Elysées, Les Six—with the exception of Honegger, too ill to be there, but including Milhaud, in a wheelchair—recently forgathered to hear some of their works played at a concert marking their thirty-fifth anniversary. None of the young Paris intellectuals attended. The enormous audience was composed of people who were young thirty-five years ago, when Les Six were *le dernier cri.* The major work played was a suite from Auric's ballet score "Phèdre," fresh and musicianly, rich, hearty, and delicious. Of all the program, it was most in the public's taste. Poulenc's cantata "Sécheresses," for choir and orchestra—written to a poem by the English poet Edward James, and supposed to trace a desert landscape by Salvador Dali—was magnificently melodic, and had, sure enough, an undulating musical line like a horizon. Milhaud's Second Symphony sounded important and scintillating but prolix. Architecturally, the big affair was Honegger's "Prélude, Fugue, et Postlude," from his "Amphion"— grandiose, savage, and melancholy. All in all, it was a melancholy concert. It recalled the wonderful first postwar Paris, when the Kaiser's Germans had just been defeated and Lenin's Russians worried no one yet.

November 24

The Parliament's present debates are only the semifinal. None of the various proposals for a united Europe are now up for ratification. They may be fought out in the new year and to judge by this past week, perhaps without there being enough of a majority either to kill the ideas forever or push them into effect. The debates have proved principally that to the French these proposals have become matters that concern the conscience (even in the case of politicians) or, at any rate, matters that arouse profoundly individual convictions, typical of the critical, individualistic French mind but, for once, all revolving about the same deep-seated belief—that it is necessary to fear a reviving Germany. The six most important veterans' associations are split exactly in two; three are resolutely hostile to the rearming of Germany in any form and three favor the proposed European Army.

Only one Parliamentary political party, the Communist Party, is of one mind about the E.D.C., and it is ready, on Moscow's orders, to vote against it en bloc. Every other party contains deputies who are for and deputies who are against the E.D.C., though the party itself may officially be against it, like the two Gaullist parties, or may officially be for it, like Bidault's M.R.P. Party. The Socialist Party is for the E.D.C. provided England comes in, too, the Socialists being fundamentally Europeanists from way back, long before the E.D.C. was ever thought of. It dominated the recent Parliamentary scene, yet it contains factions that, thanks to their characteristically severe consciences, have refused to accept the E.D.C. in its present form. As for the Radical Socialists, who include France's big businessmen, they have been so much for and so much against the E.D.C. that they have operated practically as two distinct parties. The pro-E.D.C. wing is headed by ex-Premier René Mayer, who has just said that he believed in the E.D.C. "as the indispensable pillar of the Atlantic community." The Radical Socialist anti side is led by rambunctious old Edouard Daladier, who as Premier signed the Munich pact with Hitler in 1938 and, as he grimly indicated in his recent Assembly speech, one year later lost his trust forever in any kind of Germany. In all these groups are men who belong to what should be called the Ostrich Party—usually chauvinistic or else earnest, unrealistic patriotic Frenchmen, who stick their heads in the sand so as not to see that once more a defeated and yet invincible Germany has sprung from its ashes, a Teutonic phoenix toward whom the tired Gallic cock, with his long, painful recollections of the previous fights, must now perforce take some sort of position. For the French as a whole, the E.D.C. problem is Germany—not Russia, as it is for the Americans.

Bankruptcy is something French financial experts have on their minds just now, being stoically convinced that France's démodé industry and old-fashioned agricultural system can run only another three or four years. Then, they believe, France will crack up, because of its low productivity, its high prices, and the fact that it faces modernized competition on all sides.

However—and maybe here is some good news—among both the pro- and the anti-E.D.C. French there is a distinct impression that Russia is now in no political or material position to desire anything but a continued cold war, since its European satellites, including East Germany, are in various states of unreliability and, as the Moscow newspapers themselves announced, after all these pinched postwar years the Russian people need the encouragement and reward of the higher standard of living and comfort recently offered by Malenkov.

The anti-E.D.C. French mostly think that the Americans have proved immature and often alarmingly indiscreet, and, jealous of old France's already tarnished sovereignty, rebel against any further bossing by us or lowering of France in the international hierarchy that is inherent in the E.D.C. Many pro-E.D.C. French agree with this unflattering notion of American leadership but think the world has now definitely shrunk, and that amalgamation across and around the Atlantic will be history's next instinctive step.

The French who are neutralists candidly want France and West Europe left out of Atlantic-defense treaties, so that in case there is an atomic war, Russia will not regard them as belligerents but will leave them alone, the fight for control of the world being, in the neutralists' eyes, only between the U.S.A. and the U.S.S.R. There are even some neutralist fantasists who envisage the possible future atomic war as taking place in the air between the United States and Russia, and entirely at long distance, with guided missiles being shot southwest at us by Russia and northeast by us toward the Soviet Union, both aimed in such a way as to give Europe a comfortable miss.

Both the pro- and the anti-E.D.C. French fear that Germany, out of habit, might provoke the next world war, or anyhow, once in it, would fall to conquering France, as usual, no matter what its treaties. The difference between the antis and the pros on this point is that the former believe, with all their hearts and souls, and all their hopes of France's and their own and their children's survival, that Germany can best be restrained by being kept out of any European Army, and the latter believe that Germany can best be contained in the European Army and the new European community that would go with it.

And there the two French views still stand, as they have been standing since February, 1952, the last time, up to this week of debate, that the problems were discussed in Parliament.

These may seem long-drawn-out thoughts to American readers, to many of whom the answer would appear to be "United you stand, divided you fall." But they are the thoughts that the deputies have been discussing this last week in Paris, that the newspapers have been printing, morning and night, and that the French citizens have been listening to on the radio and trying to understand in the papers, in spite of all the technical political jargon—thoughts that the French people, with an educated sense of national involvement, realize they, the Parliament, and their France as an entity must base a solemn decision on, probably in the first month of the new year. The French are aware that this decision can be a turning point in France's, Europe's, and even America's, England's, and Russia's modern history.

Victor Hugo was the first popular French modernist to talk of the practicability of a United States of Europe. It then seemed like a novelist's dream. To many French, it still does.

December 7

The Goncourt Prize, for the year's supposedly most talented novel, normally becomes the leading matter of pre-Christmas literary interest of France. This December, the interest has shifted to the discreetly unpublished leftovers of the Goncourt "Journal," the diary kept by the two famous brothers, of whom the survivor, Edmond, founded the annual fiction prize bearing the family name (except that the family name was really de Goncourt, they being aristocrats). The spotlight has centered on this part of the journal because the descendants of Alphonse and Léon Daudet—the father and son to whom Edmond willed the task of confiding it to the Bibliothèque Nationale, where it was to be kept under seal for twenty years, until 1916—recently filed a court action to prevent its publication even now. It consists of the more intimate pages of gossip written by the brothers, partly in hieroglyphics, as private comments on their literary friends. For this reason, and what with one delay and another, fifty-seven years of secrecy have shrouded it, until everyone mentioned could be presumed in his grave. Recently, these leftovers, become legendary in modern French literature, were announced for publication in 1954, to come out in an unspecified quantity of volumes, two volumes at a time, priced at fifty-three hundred francs, or fifteen dollars, a pair, in a numbered de-luxe edition sold sight unseen to subscribers only. The Daudet descendants haughtily demanded that publication be forbidden unless they were given the preposterous privilege of going through the manuscript and cutting any reference to Grandfather Alphonse that they disapproved of. Their lawyer claimed they had become alarmed when they heard that the diary described him as "hallucinated by flesh." In theory, nobody except the ten members of the Goncourt Academy, which bestows the annual prize, and their associates has seen the diary yet, but literary experts say with gratification that the phrase sounds perfectly like the malicious, refined de Goncourts, and just like lively old Alphonse, too, who wrote the novel "Sapho." What has made the Paris public laugh coldly at the Daudet family's priggishness is that the late Léon Daudet, who died in 1942, is still vividly remembered here as wielding the most scurrilous pen of all the editors of the prewar Royalist newspaper *L'Action Française,* notorious for its offensive vocabulary. Daudet's specialty was his abusive anti-Semitism, particularly against Premier Léon Blum.

The Daudets' protest also brought the terms of Edmond de Goncourt's testament into the newspapers, revealing its surprising proviso that the Goncourt Academy was annually to pay each of its members six thousand francs—twelve hundred dollars at the time—for reading and judging the novels. But the lucky winner of the Goncourt award received, and receives, only five thousand francs in prize money. To provide these sums, Edmond willed the academy the royalties on his books and those that would accrue "from the publication of my journal, which will be found after my death in my *boulle* [tortoise shell and brass] armoire . . . and this journal shall be sealed for twenty years." He had already published nine of the now classic fraternal volumes before his death in 1896. The Daudets' lawyer, in his plea, reportedly said that the excerpts were not worth printing, being inaccurate, indiscreet about several private lives, including Turgenev's, Balzac's, and Flaubert's, and full of braggadocio anecdotes of the sort that "even great literary men relate after a good dinner." Now that the Daudet court action has been dismissed and the long-sealed diary is sure to see the light of day in 1954, many Parisians—doubtless including the Daudets—can hardly wait to read it.

Today, the fiftieth of the Goncourt Academy awards—now, alas, worth the equivalent of fourteen dollars and twenty-nine cents—was given to Pierre Gascar for his novel "Le Temps des Morts." The latter is an autobiographical account of his experiences in a concentration camp, a form of life that the de Goncourts never heard of.

Though 1953 has been filled no more than any other year with conspicuous good will toward men, it has at least brought something like peace on earth to Korea, and the French hope that 1954 will bring it to Indo-China.

December 22

As the septennial Presidential election dragged on vainly through its fourth, fifth, and sixth day and night at Versailles, it began to resemble an accouchement in public, a long-drawn-out, agonizing strain, with Marianne lying in labor and everyone looking on, an indelicate spectacle of prolonged embarrassment. "PITIE POUR LA REPUBLIQUE" and "ASSEZ, MESSIEURS," newspaper headlines cried, as if the repeatedly balloting politicians, by their clumsy selfishness, were pitilessly inflicting real pain, more than could be borne, on the French body politic. It has been the worst, most weakening, and most disillusioning democratic experience France's Fourth Republic has suffered, and it has come, as usual, at the hands of the political men elected to strengthen and govern it. Only the French, with their unfortunate uninterrupted experience of little that is good from their Parliament, could have weathered it.

It all began so pleasantly. Thursday, Election Day, was the last warm, sunny day of the false spring that Paris has been enjoying at the year's end, and in Versailles's gardens finches were singing and primroses destined for April were mistakenly blooming. (That night it turned cold, and the next night it snowed slightly, as everything began to go from bad to worse.) For the Hôtel Trianon-Palace's famous septennial luncheon—traditionally served at high noon, since the voting begins at two—an impressive *tout-Paris* crowd of nearly a thousand had assembled. This is one of the few functions that carry on with the social energy of prewar days, probably because it occurs only once every seven years. The ladies were much less dressed up for the occasion than they used to be, but the crowd was as special as ever—Cabinet ministers, playwrights, scientists, big businessmen, actresses, duchesses, novelists, ambassadors, Quai d'Orsay people, wives generically, journalists, editors, and so on, plus leading French politicians, including Presidential candidates. For the three-thousand-franc *prix fixe,* the lobster was succulent, the champagne dry, the armagnac fragrant, the gossip unbridled. Some of the candidates and their wives had optimistically brought and stowed upstairs in bedrooms their top hats, tails, and evening gowns for the triumphal drive back to Paris, if they won, accompanied by the Garde Républicaine on horseback and in white breeches.

The Salle du Congrès, in the Versailles château's south wing, where the balloting takes place, is an uncharming place, rebuilt in 1875. None of us journalists had to look at it long. The press gallery, near the ceiling, consists of three steeply pitched little red velvet benches, which you clamber over at the risk of breaking your neck and which seat five spectators each. Here the several hundred assembled journalists took turns sitting for a few minutes. The sight below was unanimous, even with the President of the Congrès presiding in evening clothes in the afternoon, and, below his podium, the senators and deputies silently putting votes in the urn—the rule against harangues being so strict that when a priest deputy began to chivy the Communists, the President solemnly snapped, "Not even the voice of God can be lifted here." The only possible excitement can come when the winner is announced, which in the past was usually after the first ballot. At the present writing, the Congrès de Versailles has voted ten times in an unprecedented, scandalously selfish, ambitious, party-ridden dogfight to elect the President of the Republic—whose post symbolizes national unity and who himself is apolitical, since, once elected, he must resign from his party in order to be the arbiter of all—and still has elected nobody. In this last round, René Pleven's Union Démocratique et Socialiste de la Résistance refused to vote at all, in shocked protest at the awful way things are going. In the eighth round, fifty-three votes were listed by the ushers as *divers*—cast for diverse personalities who had nothing to do with the election, such as the Pasha of Marrakesh, Marshal Juin, novelist François Mauriac, General de Gaulle, and Marcel Boussac, the millionaire horse racer, along with twelve votes for René Laniel, brother of Premier Joseph Laniel. Had Premier Laniel received all fifty-three joker votes, he would have had a majority and been elected. Because he is an Indépendant and rich, it is natural that he should represent the Right in one of the two cleavages dominating the Versailles struggle. The European Army is the other, and even more violent, election dividing line, which was unexpectedly set up by Secretary Dulles's "agonizing reappraisal" speech, and it is less natural that Laniel, who is himself lukewarm on the European Defense Community, should now be loaded with it as the Premier who went to Bermuda, even though he was too sick there to count. It is just as ironic that Marcel-Edmond Naegelen, next-to-top candidate, who represents the Left, is a Right Wing Socialist and anti-E.D.C., although the Socialist Party favors E.D.C. (if England comes in). To Parisians, Naegelen is mostly familiar as the unlucky name signed to the pass-the-hat appeals in the Socialist newspaper *Le Populaire* to keep that staggering gazette running. Now *Le Populaire* limps behind by one lap in all the election returns, failed to announce Friday that Naegelen was then ahead, reported the ninth ballot when other papers were reporting the tenth, and remains forever one day behind the times, owing to its being printed somewhere in the provinces, where it saves money and loses the latest news. Also, the two and a half million postage stamps that so far have been imprinted with the special château cancellation "Congrès de Versailles"—always a philatelic feature of the election and particularly so this year, with a new stamp by Utrillo depicting the château gate—are still being dated December 17th, when the elections began. These two small phenomena, in which time is always a little out of whack, add to the peculiar air of destructive unreality that surrounds what is going on—and on—at Versailles, as if the Congrès out there had lost count, too.

This year's politicalization of the Presidential election may permanently change and corrupt it. In writing the Fourth Republic's constitution, the deputies made the President's powers so weak and theirs so strong that they have constantly overthrown the government, and have been forced to repair to President Vincent Auriol so often for counsel on a new Premier that he, given seven years' practice in negotiation, skillfully strengthened his office until he has become the most influential official in France. Suddenly seeing the value Auriol had created in France's only stable high office, hitherto rather ridiculed, each party last week stampeded forward to try to put its man in his shoes. Dulles's brutally candid warning to France to decide soon on E.D.C., or else, also increased the post's value in the eyes of politicians, who on each side of the fence are determined to elect as President a man who, merely by his former record, will be identified by Washington, London, and Berlin as a pro- or anti-European Army man. These are the reasons the politicians rushed in to politicalize the innocent Congrès de Versailles, which in one week of greedy maneuvering they have pulled down to the level of Parliament. At no other time in the history of this republic has public opinion expressed itself with such a vocabulary—in such words and phrases as "national catastrophe," "sinister farce," "parody," "sordid," "stupefying," "revolting," "unworthy," "indecent squatters," "lunar quarrels," "lunatics," "heartbreaking spectacle," "degradation of political spirit," and "dominating sense of humiliation."

There is an improbable rumor that Auriol himself, who is tired and old, and has had enough, will consent—if the indecent squatters pull out—to run for, and surely win, his high office again, in order to save its repute. "If he will do that for us," a young Frenchman said, "all we can say to him will be simply, *'Merci, Monsieur le Président.'*"

1954

June 18

Many people here think that if on June 4, 1953, when Pierre Mendès-France made his first bid to become Premier of France, Parliament had accepted him instead of choosing Joseph Laniel, the country would not have got into such dire straits. It was his program as would-be Premier last year that brought him forth as France's possible new political hope, the Left Wing chief of the powerful, middle-of-the-road Radical Party, the leader of the opposition against Laniel, and thus the natural politico to try to succeed him. Mendès-France's entrance into the Premiership was odd. In his speech to Parliament yesterday afternoon, he offered what he called a contract, valid for four weeks, to deliver his political promises like goods. While the Chamber sat astounded, he said, "Today is the seventeenth of June. I will meet you here before July 20th and give you an accounting. If no satisfactory solution has been found by that date, you are released from the contract, and my government and I will resign." The three priceless items he promised to deliver by then are: first, a cease-fire in the Indo-Chinese war; second, an economic-reform program for France (plus attention to a set of real political reforms for rioting Tunisia and Morocco, where, he thinks, old-hat French colonial policy is to blame); and, third, a proposal on European defense, on which a decision must be taken by the deputies before the Parliamentary vacation. Of Indo-China, he said, "If the conflict is not settled quickly, there is risk of war—of international, and even atomic, war." Before closing, he told the deputies, "I am asking for a clear answer. I am offering you a contract." In a vote held in desperation at 1 A.M. today, the contract was signed.

There is always a theatrical air in Parliament when a would-be

Premier—especially an inexperienced contender like Mendès-France—awaits his fate. On this occasion, the visitors' galleries were lively with women in summer frocks, dressed as if for a matinée; along with them were men from the embassies, French Army officers, and a solid listening bourgeoisie. At first, the would-be Premier sits alone on the Ministers' bench, since he has as yet no Ministers around to befriend him. The sense of theatre heightens when he mounts the tribune to speak his piece, like a soliloquy. Mendès-France is not a facile speaker; for a man with a big head and big shoulders, his voice is weak and uneasy. What makes his speeches is his ideas, which are hard and clear.

Mendès-France is of a Sephardic Portuguese-Jewish family. His father owns a prosperous *maison de confections pour dames* at Passy. His wife is Egyptian by birth, her family having become rich through their Egyptian chain stores, something like Woolworth's. She is handsome and a talented portrait painter. Mendès-France's career has been peculiar, distorted by the unpopularity of his strong ideas. Fresh from the University of Paris law school at the age of nineteen, he was the youngest lawyer in France; he was a deputy at twenty-five, a few months after passing the age limit required for the office; he had his first post, as Undersecretary of State for the Treasury, in Léon Blum's 1938 government. In the war, he flew a bomber and was a captain in de Gaulle's Free French forces until 1943, when the General asked him to get out of the air and become Commissaire de Finance in the Algiers provisional government. After the Liberation, he was the General's Minister of National Economy. But his ideas were too Draconian to be thought feasible, let alone to be popular, and he eventually became a sort of independent, brainy misfit in the Radical Party. He is considered to have the best head for economics of all the politicians. Since 1950, his ideas on the Indo-Chinese war have always been at least a year ahead of anybody else's, and therefore unacceptable. He has long demanded trained Vietnam officers and a firmer attitude toward conscription of native troops, and last year, as a would-be Premier, he demanded in his program a negotiated peace with Ho Chi Minh, which he will now try to obtain—"with fewer trumps than France held a year ago," he pointed out with spirit to the silent deputies.

There was a moment during the recent debates when it seemed as if he would be ruined by the Communists' insistence on giving him their hundred votes because he promised to stop what they call the dirty war. But then he insulted the whole Communist Party roundly by saying he did not want their votes counted in his possible majority. "What would our soldiers think if they learned tomorrow that their country is in part governed by men who refused to rise to their feet and honor their dead?" he asked. This reference to the Communists' recently having sullenly sat while the Assembly rose to honor those lost in Indo-China probably saved him.

Even though Mendès-France's Draconian ideas have never been popular, among the French citizens he is certainly the popular choice. They want to give his ideas a chance, at last, to function.

The outstanding result of their recent bitter experiences, the French now say, was that in the fall of Dienbienphu they discovered their loneliness. The main outcome of this belated knowledge of their isolation is that during the past six weeks, according to reports submitted to the government by prefects all over France, about sixty-five per cent of provincial opinion has come to favor E.D.C. in some form, apparently even including the European Army. Country parliamentarians grandly tend to become Parisians as soon as they are elected, and they, plus all the special political, nationalistic, and anti-German sentiments that are native to France's capital, have made Paris the center of the anti-E.D.C. movement over the past two years. But the hardheaded French countrypeople, more afraid of Moscow than of Bonn, now realistically accept the premise that West Germany, at present a powerful vacuum, will be armed eventually by the United States if it is not armed by E.D.C. Above all, they are convinced that in a world of mighty giants like the Union of Soviet Socialist Republics and the United States of America there must be something like a United States of Europe, so that Europeans, en bloc, can make themselves as big as possible in a new shape. Far behind the military crisis in Indo-China is still the real French historical crisis of what to do—and at once—about European defense. Georges Bidault's M.R.P. Party, the only one wholly sworn to E.D.C., abstained from voting on Mendès-France's Premiership, because he is distinctly not sworn to E.D.C. in its present form. Prior to the voting, it is said, the M.R.P. hoped to obtain new Parliamentary elections. The M.R.P. idea was that with a new set of deputies, truly representing the provinces (which really mean France, as Paris does not), a Parliamentary vote could be taken during the summer that would offer France her only chance of positive decision about her destiny—her chance of identifying herself as leader in a union of Europe, which was her pragmatic invention in the first place. To pro-E.D.C. Americans here in Paris, the whole project sounded inspired and impossible.

June 30

The victory of Pierre Mendès-France in becoming Premier, if only for the four weeks he gave himself to perform what are beginning to look like obtainable miracles, has had an extraor-dinary effect on the country. On all sides there exists "a more or less general hopefulness, even though it may be discolored by skepti-cism," as one newspaper phrased it. Informal polls have been hastily taken to find out what the citizens feel and think about their new leader, as if for once their reactions mattered. The findings are illuminating. The *populo,* or poorer class, and the topmost educated class are the most enthusiastic about him, the middle class the least. There is more incredulity among businessmen and industrialists about his chance of fulfilling his month's promises than there is among the intellectual and technical cadres; high-profit and low-output French business and industry have been screaming for reform, mostly on taxation, but they already suspect that in his economic-reform program he may try to reform them, too, and, naturally, believe that he cannot succeed. On the other hand, in a wave of curiosity, a recent conference of thirty highly educated technical experts in Paris industry took a vote on its feeling for the Premier, and the returns were twenty-seven for, only three against. And a recent reunion of *lycée* and university professors recorded a nearly unanimous outburst of confidence in the Premier. The students at the Ecole Normale Supérieure and at the Polytechnique, who are the cream of young French brains, are mostly for him, because they think he has an excellent mind, clear formulations, no tergiversations, and is honest, and because he is only forty-seven and full of sap, and has chosen six men under forty as Cabinet Ministers and State Secretaries. The Army brass tends to be for him, because he named as Minister of Defense General Pierre Koenig, who fought the Laniel Defense Ministry over the poor pay of combat officers in Indo-China and also over its conduct of the war. The support of Mendès-France that is closest to bedrock is probably that of the *fonctionnaires,* or civil-service employees, who keep France running for whatever government comes along—and even when there is no government at all—and who have a correspondingly poor view of politicians generally. Lower and middling *fonctionnaires* are out-and-out for him as the exceptional politician, and among the mature chiefs, men of importance in the state machinery, are many who might be kindled by him into hope for a renewed, cleansed France —a France they could more gladly serve.

Mendès-France's Parliamentary base is certainly shaky. The political parties represented by the men now in his Cabinet have a total of only about two hundred and seventy-five Assembly votes—and the parties involved would not, any more than his own Radical Party, necessarily support him generously. He even has two fence jumpers from Bidault's M.R.P. Party, which gave its members strict instructions against taking office with this newcomer, the first Premier since the war to put the M.R.P., as well as Bidault, after his long, faithful, uneasy seasons as Foreign Minister, on the shelf of the opposition. The M.R.P.'s votes against him will be triply bitter.

Mendès-France's contract time is nearly half up. He has already created a Cabinet post new to French history—that of Minister for Moroccan and Tunisian Affairs—as the first step in carrying out his plan for new relations with those riotously inflamed protectorates. He has done what in his *Premier-désigné* speech in Parliament he warned the French Communists he would do whether his cease-fire offer ended the war or not—ordered reinforcements to be made ready for the Far East expeditionary force. He has called both on prior governments and on his own to present to the Assembly early in July definite proposals for reforming the constitution. His first fiscal-reform project, just announced, aims at strengthening and enforcing last year's halfhearted law taxing manufacturers' markups, which, if made to work, would enormously bolster the national economy. Though some of the French, especially the wealthy, feared unortho-dox novelty reforms from him, he seems so far merely to be reforming the weak or inapplicable reforms other men passed as lip service to necessity or in order to keep previous reformers quiet.

The first, and most compelling, item on Mendès-France's program—to obtain a cease-fire—now looks closer, since, to every-one's surprise, Sir Winston and the President have announced that the United Kingdom and the United States will press forward with their security plans for Southeast Asia whether or not France negotiates an acceptable agreement in Geneva. This is probably the best incentive the Anglo-Saxons have yet given France in her efforts to get such an agreement at the peace conference. As France waits for the Indo-China project—physically the most vital issue to the French people, as war and truce issues always are—to be finished one way or the other, the great and almost philosophical French issue of the European Defense Community is close to boiling. It is the most nearly insurmountable problem in Mendès-France's program, because it will have to be settled by the French themselves, now and finally and at last. After twenty-five months of waiting, Belgium's Henri Spaak, speaking for the exasperated Benelux group, and Chancellor Adenauer, speaking for West Germany, this week began bringing pressure on the Premier, and the American Ambassador, C. Douglas Dillon, transmitted to him Washington's urgent demand that France ratify at least *a* European defense community. In an effort to get a compromise plan that French Parliamentary partisans of E.D.C. and those who are its enemies would both accept, the Premier had already set General Koenig, who is anti-E.D.C., and Maurice Bourgès-Maunoury, the Minister of Industry and Commerce, who is for it, to working on it together. Between this compromise E.D.C., which doubtless no one but the French would accept, and the old, original E.D.C., which other Atlantic nations have long since accepted, Parliament must finally make up its mind during the last week of the Premier's contract—and perhaps, of course, of his Premiership.

Mendès-France is nothing of an athlete in appearance, and nobody dreamed that he had the gift of drawing on himself the vast public eye as a politician. But as Premier he is like a runner running a unique, exciting political hurdle race while the citizens of France watch him closely, lap after lap. As far as real racing is concerned, this has been the high point of the horse-racing season in Paris, with well-dressed women, fine, sensitive animals, thin jockeys, and great crowds at Longchamp, Auteuil, Enghien, and St.-Cloud. But the Premier's race against time is the strangest historical contest anybody has ever seen run here, and France has her eyes on it.

July 14

The precipitate closing of the remarkable show of early (1900–14) Picasso paintings at the Communist Maison de la

Pensée Française, which occurred in the face of a writ for the recovery of stolen family property launched against the precious thirty-seven canvases from Moscow and Leningrad museums, has been relished here as one of the oddest happenings in all the eccentric, violent annals of modern French art. The writ was obtained by Mme. Irène Stchoukine de Keller, whose father, the Czarist industrialist Sergei Stchoukine, was the greatest Moscow collector of Picasso's young works. According to the Communist Party paper *Humanité,* which took a haughty tone, the Stchoukine collection "became nationalized property in 1918," and the writ was simply *"une prétention vexatoire et fantaisiste,* whose only result has been to deprive the French public of the sight of these works, sent for its enjoyment by the Soviet museums." *Humanité* pointed out that the lady—who indeed seemed to know little about her father's collection except that she naturally wanted to get her hands on it—even included in her list of stolen items in the exhibition the Cubist gem called "La Femme à la Guitare," which actually belongs to the late Gertrude Stein's collection. The paper also humorlessly quoted the celebrated Picasso art merchant Henry Kahnweiler as saying that Stchoukine, who died here in poverty in the middle nineteen-thirties, told him he had always intended to give his pictures to the Moscow museum, "only it happened a little too soon."

The court to which the claim of theft, or "seizure without indemnity," was presented immediately ordered that "everything should remain in place"—an order that the Soviet Embassy defied within the hour by whisking the pictures off to some safe, secret spot, just in case the judge should hold the claim valid and sequester them. The pictures were previously removed from sight in Russia, too, during the years when Picasso's art fell under the Stalinite cultural ban as cosmopolitan decadence; nobody here can say whether, under Malenkov's apparently less strict regime, the pictures had been declared healthy again and rehung in the Soviet museums. If, as the Paris correspondent of the Manchester *Guardian* succinctly commented, "they are in fact only shown when lent abroad, but cannot be lent abroad because of a doubt about legal ownership," these early Picassos are in a bad way. Picasso himself, who was attending a bullfight in Nîmes the day his canvases disappeared from the Paris show, said, "I am sorry. I would have loved to see my paintings again." He added that he did not think Stchoukine's daughter had a right to do what she did, remarking, "What if the Comte de Paris took it into his head someday to lay claim to Versailles?" Now that the court, as was historically inevitable, has thrown out the claim, there is no reason the Soviet Embassy should not bring the Picassos out of hiding and let the exhibition continue to be a major Paris summer art pleasure through September, as scheduled. However, nobody expects Russians to be reasonable.

A talented, short, and certainly unusual first novel by an eighteen-year-old authoress has become required vacation reading here, after having won last May's Prix des Critiques. The novel is called "Bonjour Tristesse" and is signed Françoise Sagan, a nom de plume devised in honor of Marcel Proust's Princesse de Sagan. The young writer's name is Françoise Quoirez. The daughter of well-off Paris parents, she was born in 1935, lived with her family in Lyon during the war, at seventeen entered the Sorbonne, failed to pass last July, and during the month of August—"having nothing to do," she says—wrote her first book. No matter how poor a student she may be, she is a born writer. Her style is being compared to Raymond Radiguet's. It has the same youthful, swift, unaffected manner of setting down characters and scenes clearly, as if written on glass, in the simple, permanent, eighteenth-century classic French vocabulary. Her story also has Radiguet's untroubled artistic amorality. It deals with a girl of her age, who is finally freed from boarding schools to live in unchecked enjoyment with her youngish widowed father, an affectionate, civilized lightweight immoralist whose life events are composed of rather commonplace passing mistresses, picked up in his easy Champs-Elysées business world. The crux of the story, which takes place in the Midi during the father's and daughter's summer holiday *à trois* with an especially blowzy mistress, is the arrival there of the dead mother's best friend—an intelligent, forceful Parisienne, who not only falls in love with the father and easily ousts the mistress but plans to marry him and save him, at forty, from his shapeless, tarnishing life and worse future, and to bring up the girl *comme il faut*. The resultant tragedy, in which the girl takes her first lover, is the struggle between two species—between the daughter, who instinctively fights to save her dronelike father and herself for the hedonism their natures intended them for, and *la femme forte*, whose superior, logical plan for salvation the girl intellectually recognizes but destroys. Defeated, the woman is killed, or kills herself, when her car plunges off the Corniche, leaving the girl and the father victorious and lost. Among the exceptional appreciations that "Bonjour Tristesse" has evoked is a moaning editorial by François Mauriac, who deplored the fact that a prize should have been given to "this cruel book, whose literary merit shines out from the first page and is indisputable," called the author "a charming little monster," and bewailed today's *"dévergondage de l'adolescence féminine."* Testy words on the subject of literary moral values followed this in some papers by critics who pitilessly analyzed Mauriac's new novel, "L'Agneau," as containing an extra helping of the sordid goings on, always without pleasure, which his provincial Christian characters are invariably involved in on their ultimate way to Heaven.

July 28

As a small proof of the finality of what has just been arranged in Indo-China, the aged King of Laos has arrived for an extended stay in France, where he will take the waters at Vittel, and Bao Dai, chief of state of Vietnam, is already taking the waters at Vichy.

There can be no doubt that Premier Mendès-France has become practically a national hero in the month it took him to fulfill his fantastic promise of an Indo-China cease-fire, which the French have generally called anguishing good news. He is now regarded as having done three major things concurrently in four weeks—ended a colonial war that might have become a Third World War; offered postwar France another chance, perhaps cheap at the price, at realistic political and financial salvation; and, along with Anthony Eden, revived the Entente Cordiale of that European-minded monarch Edward VII, substituting London across the narrow Channel for Washington across the vast Atlantic.

August 11

"Funérailles de Colette," her bare name printed in heavy black, with no other identification—only that of her fame—was the phrase used on the official invitation to her public outdoor funeral, accorded to her by the French state. It was held in the Cour d'Honneur du Palais-Royal, adjoining the palatial tree-filled and sculpture-filled Palais-Royal Garden, at the opposite end of which she had lived for many years, and where she had died suddenly, and painlessly, in the evening, after a small sip of champagne. It was appropriate that she died during a brief burst of magnificent rural weather here in Paris, on one of this summer's rare days of provincial heat, yet when autumnal colors were already on her garden trees, which themselves looked provincial rather than metropolitan; as a Burgundian-born countrywoman, she was sensitive to the point of genius in her pleasure in nature, and her written reports on it are unique in French literature, where she left that intimate record of her observant, penetrating love of creatures, blossoms, and foliage, of fields, woods, and seasons. She was aged eighty-one, had been immobilized in late years with arthritis, and had been a writer since 1900.

Colette's state obsequies were the only honor of their kind to be given a French woman writer in the history of all the four republics. She had experienced triumphs in her life, and now she lay in state on a great catafalque loosely covered with tricolor silk, which the wind billowed, and her glittering cross of a Grand Officer of the Legion of Honor and its scarlet ribbon were attached to a black velvet pillow that leaned against her coffin. Facing the catafalque sat her daughter, Mme. Colette de Jouvenel; Pauline, her faithful maid for a quarter century; and her last husband, Maurice Goudeket. Flanking them were members of the Goncourt Academy and other French notables. Across the court stretched a hedge of floral offerings, with pale-blue and dark-blue gladioli predominant. (Blue was her favorite color, and she always wrote on blue paper.) There was a wreath from the French Parliament; one from the city of Lyon and its Mayor, Edouard Herriot, an old friend of hers; a garland of roses with ribbons of the Belgian colors, marked "Elisabeth," from the King's grandmother; a huge sheaf of lilies from the Association des Music-Halls et Cirques, since in her difficult young days she was a music-hall dancer and mime; and a big country bouquet of dahlias from her "Compatriotes de St.-Sauveur-en-Puisaye," the Yonne village where she was born. Fenced off behind the catafalque and stretching toward Colette's corner of the garden, stood, very quietly, the great overflow Parisian crowd, her anonymous admirers.

As a writer and personality, Colette came with time to sum up for the French certain French essences, as in her love for excellent foods and wines, her perfect feeling for the written or spoken French phrase, and her literary sense *de l'époque,* whether the epoch was that of aging cocottes or of nubile provincial schoolgirls. In her writings, she even transferred to French readers her love for nature and for animals, domestic or roaming country hills. Tastes such as these had been absent—in an almost too civilized omission—from French literature until she brought her Burgundian girlhood to Paris and put her rural sensibilities on paper. She had a sensual harvest fullness in her writing and in her person, and a profound feeling for love springing up or being mowed down. She was indifferent to moral calculations, and composed her novels and stories with her informed instincts rather than by building any formal plot, and thus left behind her pages that are uniquely feminine and the work of a master, in which the perceptions of her five senses were made permanent through her genius for the exactitudes of the French language. She was indeed a writer who was "pagan, sensuous, Dionysiac," as the Ministre de l'Education Nationale, Jean Berthoin, loudly declaimed with admiration in his funeral address at her bier.

However extraordinary these words seemed, from such a source on such an occasion, there was much more just as extraordinary that was said, and also written, in the course of tributes and obituaries, for her death ranked as the greatest literary loss since that of André Gide. The director of *Le Figaro,* Pierre Brisson, wrote on the front page of that ultra-Catholic daily, "There was something androgynous in her; she was masculine in her initiative and authority, and a woman by her intuition and assent; she was both vigorous and a dreamer." Germaine Beaumont, the novelist, who had since childhood been intimately acquainted with Colette (who was Germaine Beaumont's mother's best friend), wrote in *Les Nouvelles Littéraires,* "How alive she was, to the degree of making others seem moribund! She had more sensitive tissues . . . and more aptitude than others for observing, seizing, holding. She who learned so well how to remain immobile, drawing repose over her like a coverlet, sometimes reminded one of those marine creatures that breathe, nourish themselves, and even circulate by means of their fringe of sensitivity." As his farewell, the poet Louis Aragon wrote her a fine, rousing poem of adieu based on a quotation from Tasso.

The two novels the critics cited most often this week as the quintessential Colette were "Claudine à l'Ecole," which was the first book she wrote, and which was published not under her own name at all but under that of her first husband (soon afterward divorced), the insouciant journalist Henry Gauthier-Villars, known as Willy; and "Chéri," regarded as her masterpiece, which tells the story of a tarnished, aging cocotte and her tragic, belated love for a young man. (Colette acted the role of the cocotte, Léa, in a stage version produced in Paris in the mid-nineteen-twenties.) "Le Blé en Herbe" was one of the great public favorites, and "Gigi," one of her last works, was popular as a novelette, as a film, and as a play on both sides of the Atlantic.

Colette's funeral, *aux frais de l'Etat*—the highest posthumous honor that can be granted to a citizen—viewed alongside the reaction of the Catholic Church to her death, illustrates in a remarkably clear modern fashion the still existing separation of Church and State. Colette's second marriage was to Henry de Jouvenel, editor-in-chief of *Le Matin,* from whom she was also divorced. She was thus a twice-divorced *libre penseur,* as well as a noncommunicant of the Church at the time of her death, although she had been born a Catholic. A request, made by friends of her family, that she be given a final burial service at the Eglise St. Roch was refused by Cardinal Feltin, the Archbishop of Paris, and this refusal set off a great deal of comment and argument here. In the weekly *Figaro Littéraire,* the English Catholic convert Graham Greene contributed a front-page open letter, in French, addressed to *"Son Eminence le Cardinal-Archevêque de Paris,"* in which he took him to task for his decision that no priest should offer public prayers at Colette's funeral. He wrote, in part, "Would you have invoked the same reasons if she had been less illustrious? . . . Are two civil marriages so unpardonable? The lives of certain of our saints offer worse examples . . . Your Eminence, without knowing it, has given the impression that the Church pursues an error even beyond the deathbed . . . To non-Catholics it could seem that the Church itself lacked charity . . . Of course, Catholics, upon reflection, can decide that the voice of an archbishop is not necessarily the voice of the Church." He ended the letter, *"avec mon humble respect pour la Pourpre Sacrée."* As literary circles noted, there was an analogy between Colette's burial and that of George Sand, over the past hundred years the other vastly popular French woman writer and public literary figure, who died three years after Colette was born, who was also a free-thinker and led her own life, and for whom the state also proposed a national funeral. George Sand's daughter, to the indignation of Flaubert and other friends, preferred a private religious service, which was permitted by the archdiocese. The Church took a stronger stand on Colette, perhaps using her fame to magnify its warning to other intellectuals of little or no faith.

Early this year, Colette spent a holiday at one of the Mediterranean resorts, where she was carried around in the sunshine in a sedan chair. At home in Paris, since she was practically bedridden the last several years, she sensibly had her bed placed on a platform by a window overlooking the Palais-Royal Garden, and there she continued to write—surrounded by her cats, her famous multicolored collection of glass paperweights, her books and mementos—and could look out on the active life of children and passersby under the trees below. A plaque is to be placed on the garden wall of her home, which will read, *"Ici vécut, ici mourut Colette, dont l'œuvre est une fenêtre grande ouverte sur la vie* [Here lived, here died Colette, whose work is a window wide-open on life]."

Mendès-France is still working up momentum. On the occasion of his meeting last Tuesday with Parliament, one of the few he has had since becoming Premier, he emerged from the Assembly's two sessions, morning and afternoon—which, in fact, dragged on until almost midnight—with the exceptional record of two votes of confidence won in one day. The morning vote, giving him special powers until next March 31st to carry out his sweeping economic reforms, followed the most applauded portion of his speech to the deputies, in which he declared, "France has been plunged into a deep sleep filled with nostalgic dreams and nightmares that have fed on the black future. We must wake her up."

The afternoon meeting, where the Premier's intention had been merely to ask that the scheduled debate on the grave North African problem be postponed until after the August 24th debate on the even graver problem of E.D.C., unexpectedly developed into a political duel on the Tunisian situation, and this swelled into the most tumultuous session of the Assembly in years. The fight was started by Deputy Léon Martinaud-Déplat, who, ironically, is chief of Mendès-France's own party, the Radicals. The deputy, long opposed to Mendès-France's ideas in general and to his project for Tunisian home rule in particular, accused the Premier of negotiating with enemy terrorists in giving the Neo-Destour nationalists representation in the government. The Premier's counterattack was adequately rich, since he had all the preceding governments' North African failures to draw on in defense of his own reform plans; he cited their repeated promises to Tunisians of autonomous rule, made every year since the Liberation and never kept—with the recent appalling North African violence and bloodshed as the dreadfully logical result, to his mind, of long years of French bad faith and Tunisian frustration. The Mediterranean colonial problem, and how to manage what is left in that area of France's shrinking and rioting empire after the loss of half of Indo-China, is a topic to inflame politicians, especially those who have become wise after the event, and as the two men argued and riposted, the hemicycle alternately was thrown into an uproar, when the deputies joined in pell-mell with their own exhortations, and was reduced to tense silence, while everybody tried hard to miss none of the two antagonists' finer political points. During the general outbursts, the Assembly President, André Le Troquer, high up on his private platform, made vain attempts to restore calm, whacking his desk with the schoolmaster's ruler that serves him as a gavel, and once even ringing his big bell, which is used only for moments of serious disorder. In the crammed visitors' gallery, people in the front rows were hanging over the railing in excitement, and those at the back were standing to see and hear.

At the start of the duel, there was a moment of trepidation for the whole Assembly, and particularly for Mendès-France's followers, when he broke in on his adversary's speech to say calmly that if he did not obtain a vote of confidence on his North African policies, he would resign. Parliament could not risk the onus of causing him to fall just now, if only because of his popularity with the people as the first man in a long while who has tried to get anything and everything done. So, with the second session's vote, he received his extra victory of the day. His audacity, his honesty, his skill at debate, and his drastic realism turned the duel into one more dramatic triumph for him—naturally, a far lesser event than the winning of his month's race on the Indo-China cease-fire, and a lesser one, too, than his surprise flight to Tunisia to call on and talk with the Bey, which was something no previous Premier had ever done during any of the many troubled times there. But all three incidents have a similar quality of vigor about them that catches French eyes and imaginations and that so far has made for a remarkable success with the citizens. For him, everything perhaps now hangs on the result of the Assembly's debate on E.D.C. during the next fortnight, and much that affects the whole world, too, will there finally become a matter of history.

September 1

The greatest tribute to the magnitude of the European Defense Community question is that the partisanship it aroused cut France in two—the people, as well as the Parliament. The three-day weekend discussion of it in the Assembly before it was symbolically voted down raised individual convictions and hopes, divergent patriotisms, angers, and passions, and divisions between men and men, rather than between parties only, to an altitude of tension reached by no other issue since the war. Reams of journalistic reporting appeared on the factual results of these feelings, but only *Le Monde* mentioned *la mystique* of E.D.C. as rational news. It said that this emotion for and against E.D.C. sliced across all accustomed separations, so that a pro-E.D.C. Frenchman felt closer to a German who was for the treaty than to a Frenchman who was against it; so that a Communist who was against it (as all the Communists were) felt closer to his boss or to an Army general, if his boss or the general was hostile to it, than to a fellow-workman who wanted E.D.C. ratified. What the French people *en masse* felt about E.D.C., history will unfortunately never know, since no plebiscite was ever taken. On general principles, Frenchmen have all along been against any rearming of Germany, because of their recent experiences in the war and the Occupation and because of the German wars that fell on their fathers and grandfathers. But it would seem that before the debate opened, slightly more than half the French citizens were in favor of German rearmament under E.D.C. if rearmament was inevitable, maybe because they could not comprehend the reputedly dangerous pitfalls for France in the complex treaty—pitfalls that gave the expert anti-E.D.C. parliamentarians their influential talking points before the other deputies and the public galleries. It is unfortunate, too, that history must record that, in a parliament lately so notorious for endless debates, everyone was not allowed to have his say on this dominantly affecting matter; that the vote was not a direct and dignified one on the specific question "Are you for E.D.C. or against it?," which, after two years of tergiversation, would have steadied France's reputation for clarifying a final decision, but was a tangential and procedural one, pushed through by the anti-E.D.C. forces as if in a quick panic. The Defense Community had the air of being hastily strangled, instead of being allowed, in accordance with proper democratic processes, to perish slowly—duly done to death by words from those who did not approve.

On Monday morning, by which time the Premier had already accepted the idea of returning to Brussels, the Cédistes, in an ill-advised maneuver against the Anticédistes, brought up their *motion préjudicielle* again, even though it was legally bound to be heard, if ever, only after the *préalable,* which had been introduced first. As an anti-E.D.C. riposte, and because a *préalable* can be spoken for only by its signer, old, infirm, Anticédiste Edouard Herriot, honorary President of the Assembly, where he actively reigned for a genera- tion, was made a co-signer and was sent in alone to speak as the Anticédistes' trump card. Crippled by illness and his eighty-two years, and unable, with his swollen limbs, to mount the steps to the speaker's podium, above which he used to preside, he spoke seated, using a microphone installed for him on the Radical Socialist bench. His voice still boomed, but his facts had become weak; he muddled treaty clauses and figures and ignored shouted corrections. And at the end, for his great prestige of age and service, and for his flaming patriotism for a sovereign France, still intact, he was acclaimed with an ovation. He said, in part, "On the threshold of death, let me tell you that E.D.C. is the end of France, that it is a step forward for Germany and a step back for France, that when a people no longer runs its army, as France would not under E.D.C., it no longer runs its own diplomacy." Part of his task, he went on, was to explain to the Americans how wrong they were in thinking that E.D.C. was still good for France. *"La C.E.D. est une aventure, ne la faites pas!"* he ended, amid salvos of applause. He had been made the *vox humana* of the Anticédistes, and his speech was highly influential. By seven o'clock in the evening the European Defense Community was dead. While the Communists, who had won a victory with their ninety-five votes, rose and sang the "Marseillaise," joined by some Socialists and Anticédiste de Gaullistes, the defeated E.D.C. mem- bers shouted at them, in anguish and fury, "Back to Moscow!" The veteran Cédiste Paul Reynaud, his intelligent, small face drawn with emotion, ran up onto the speaker's podium and cried, "For the first time since there has been a French Parliament, a treaty has been voted down without its author or its signer being allowed to speak in its defense!" With this bitter obituary, the Communauté Européenne de Défense was buried in France.

The position of Mendès-France in all this remains obscure, even to his close friends (or so some of them say), as well as to his political backers, who risked much in their enthusiasm for him, and to his multitudes of French admirers. In his Indo-Chinese cease-fire transactions, he had been coolly successful in dealing at first hand with exigent foreign diplomats at Geneva, and his energetic *brio* had been deemed magnificent. Then, his negotiations with his own Assembly on the thorny, bloody question of reforms in Tunisia were a real political triumph, such as no other French postwar leader had ever expected, or even attempted, to achieve, and a brilliant example of his ability to manage and lead suspicious French politicians. But, as he told the Assembly in his recent foreign-policy speech, with the curious bold candor that is one of his appeals to French citizens, his Brussels experience was at times a humiliation. He was accused by the Cédistes of having been too hard with his confreres there to do E.D.C. any good. A French colleague of his at Brussels said, "He seems too intelligent a man to be an apt negotiator." One of his slogans, which he used as the title of a book of speeches, is *"Gouverner, C'est Choisir"*—"To Govern Is to Choose." A repeated caustic criticism in the Assembly last week was that while he urged everyone else to choose sides on E.D.C., he himself chose neither side. He also kept his government divorced from any responsibility for E.D.C., which meant that his government could not fall on a vote of confidence. Indeed, some of the bitterest criticism was provoked by this fact—that he took neither responsibility nor a side. He was perhaps too logical in thinking it his duty impersonally to force a decision on the treaty. Mendès-France's attitude was, of course, an oddly transcendental one for a French Premier. In his strict refusal to reveal his attitude toward E.D.C., lest he influence the Assembly in what the French called *cette question de conscience,* he acted with the detachment of a constitutional monarch—rather as Queen Elizabeth might conduct herself.

The tragedy of Mendès-France is that he wanted to settle the question of E.D.C. one way or the other, in order to release France from her immobility and start her on her liberated modern way. He wanted to inaugurate a French New Deal that would save French capitalism and satisfy French workers. He wanted a positive majority in the Assembly without the Communists, whom he wished to treat as if they were on an island; when he was invested, he scornfully told them that he would not count their votes in figuring his majority. In the recent E.D.C. vote, it was the Communists who supplied the majority that carried what was supposed to be his side. He had remained so enigmatic as to provide his own fatal dilemma. The intellectualism of Mendès-France and the gerontocracy spoken for by old Herriot, both so representative of France, seem to have set France floating loose and alone at last.

September 15

"The Alice B. Toklas Cookbook"—it seems that the "B" stands for Babette, which was the name of her French godmother—was, as Miss Toklas circumstantially states in her preface, lately written in Paris during a bout of jaundice, "when good foods hover in the invalid's memory," and will soon be published in New York. It constitutes delicious reading, in the opinion of Paris friends of hers who have seen the manuscript and who have, at her memorable lunches in the famous Rue Christine flat, eaten some of her gastronomic treats, for which she gives three hundred and fifty-eight recipes in the book. As might be expected, the Toklas opus is an unusual and often very droll cookbook, firm in its statement of essential eighteenth-century French cooking principles, and interlarded with comments, as dry as her favorite Chablis, on the kinds and classes of French who eat this or that, and in what type of bourgeois apartment or country château. It is also seasoned with pungent recollections of Miss Gertrude Stein, and contains as well a kind of palate history of war foods in France, which the two ladies knew in their two wars. In the first war, they munched luscious Alsatian cookies (the recipe for Alsatian cookies is given) while visiting the heroic ruins of Verdun, and they ate classic regional dishes (recipes also given) at local inns while driving their old Ford from Lorraine to the Spanish border to deliver medical supplies for the Red Cross. In the recent war, the degree of their extremity while living in their country house in Bilignin, near the Swiss border, is made clear in a chapter called "Murder in the Kitchen." Here Miss

Toklas practiced her first assassination when her fishman kindly gave her a live carp, to which she had to take the knife, feeling weak afterward (a recipe follows for carp and homemade noodles, plus her editorial assurance that "these noodles are very delicate"); later occurred the grimmer business of smothering pigeons, since the French think neck-wringing brutal, besides being a waste of the blood essential to the rich taste of the fowl. (A first-rate recipe follows for braised pigeon on croutons.) There is also the instance of the demise of the two ladies' Barbary duck, which was chased one day by a neighboring farm dog. The household's quondam peasant cook decided that it was about to die of heart failure, so she gave it three tablespoonfuls of brandy to swallow, "to add flavor," and then killed it—and how did Madame wish it to be cooked? "Surprised at the turn the affair had taken," Miss Toklas writes, "I answered feebly, 'With orange sauce.'" (Properly in sequence comes a recipe for duck and oranges.) Her Liberation fruitcake (recipe given) was concocted with the help of a few dried orange peels saved in a jar for over four years against the great day when, to everyone's astonishment, the American Army really did march into unimportant Bilignin.

As a sample of her domestic recollections of Miss Stein, Miss Toklas remarks, in connection with gooseberry-gathering and jelly-making at Bilignin, "Wasps, hornets, and bees rarely sting me, though my work with them has always been aggressive. Gertrude Stein did not care for any of them, nor for spiders, centipedes, and bats. She had no violent feeling for them out-of-doors, but in the house she would call for aid." The book also contains strict rules to be observed in good cooking, which "permits of no exaggerations," as well as thoughtful advice, such as that delicate sauces should be stirred with a wooden spoon and not a metal one, which can tinge the flavor, and that cream should be added to sauces only at the last moment, in time to heat but not to boil, and "should be tilted in the pan, not stirred." It is a book of character, fine food, and tasty human observation. Handsome decorative drawings by Sir Francis Rose head the chapters. Miss Toklas has a collection of nearly a hundred authoritative volumes on refined cooking. Her own volume can properly adorn her library.

The death of André Derain last week at the age of seventy-four, as the result of being ignobly run into by an automobile while riding his bicycle two months ago, was sad, because what he had been painting for the past twenty-five years made it difficult to recall how great he had started to be when young. In the 1905 Salon d'Automne, he showed with the Fauves, led by Matisse, and his wild-beast gorgeous colors—scarlet cabbage fields and a purple Seine—were second to none, not even those of Vlaminck, whom he had met in a railroad accident on a Paris suburban line, whereupon they both started painting in the new manner. Their paintings were like a paintpot thrown in your face, one critic said in shock. Derain later spent several summers with Picasso on the Mediterranean, became momentarily Cubist, and then went into a classical, rather noble formality of line, where he remained, by and large, until the end of his days. Recently, a court order was issued preventing him from working on or altering any of his recent paintings, since his wife, from whom he was separated, thought he might ruin them and thus diminish their sales value after his death. It was only what he painted forty or fifty years ago that today would fetch great, sincere high prices.

October 13

It is not only the size of the victory that Premier Mendès-France has just won—the Assembly's 350–113 vote of confidence in him and, by implication, in the London Conference formula, with which apparently only about a hundred and fifty deputies are actually satisfied—but the clarity, timing, and candor he constructed his victory on that still astonish the politicians who, perforce, gave it to him. It is true that he was aided by the frankness of Sir Winston's neighborly back-fence warning that there could be no question of further French fiddling with the London fundamentals, which made it clear to the deputies here that they would have to come up to scratch on the question of the fate of the Atlantic Alliance. Ironically, Molotov, in East Berlin, involuntarily aided Mendès-France by suggesting Big Four talks on disarmament, which gave the Premier, in the midst of fighting for controlled German rearmament, the chance to answer right back to him and to certain worried deputies. Yes, indeed, said the Premier, he would willingly take up such talks with the Russians, for which there was plenty of time, since the London accord—if and when finally ratified —could not operate for maybe four years anyhow.

As for Mendès-France's local difficulties, it was his raising of minimum wages a few francs an hour last week, with an April date set for further serious wage talks, that finally gave him the Socialist Party vote—absolutely essential to his majority. On the E.D.C. vote, the Socialists had bitterly split in half; at their weekend Party caucus to determine how they should vote on the London negotiations, they had labored up to their ears in argument over ideals, and the intellectual importance of remaining the Opposition party, until Mendès-France telephoned Socialist chief Guy Mollet (and let the press know about it) and realistically suggested that perhaps the Socialists might even take part in his government, which, after all, aims at being a sort of French New Deal. With French elections coming up next spring, and with Mendès-France's program of better wages, financial reforms, and action against the powerful alcohol lobby and its privileged subsidies already under way, it seems incredible that it took the Socialists another day of argument to decide to climb on the band wagon and to begin by voting en bloc for the Premier on Tuesday. Because of their dissatisfaction with the narrow social programs of previous governments, it has been the Socialists who have repeatedly overthrown them; if the Socialists should really enter the present government, they not only would look important to their working-class constituents but could act importantly by supplying this government with a regular majority, the usual lack of which is the curse of the splintered French Parliamentary system. As for the parties that still either love E.D.C. or loathe the idea of giving Germans a gun under any conditions, or that personally dislike Mendès-France and all his works, the Premier's skillful demand for a confidence vote only on the continuance of his London negotiations would clearly have loaded them—had they refused it—with the onus of having left France standing alone in history and, moreover, standing temporarily without even a government. That the hostile M.R.P. party of Georges Bidault largely abstained from voting at all, rather than vote against Mendès-France, and that the ex-Gaullist Social Republicans, who are also his enemies, largely voted for him, were unexpectedly meek replies.

In this recent victory, Mendès-France, who a few days before had been thought doomed to defeat, proved himself a master tactician. In the public mind, he took his risks openly, as part of his profession, but never gambled; he talked clearly, explained simply, promised nothing more than to do his best, and was in a hurry, as usual. Because of this, the French citizenry were able to understand for once what was going on in a vitally important Parliamentary decision, and unquestionably they comprehended the results. "He has created a political style which, if it pleases God, will survive him," piously commented François Mauriac, probably his most fervent literary admirer, in this week's *Express,* adding, "It would seem to me difficult for the successors of Pierre Mendès-France, even the mediocre and stupid ones, to go back to using the old political concoctions to govern by."

Amid the various uncertainties in France, one thing is sure: The French are seeing more flying saucers than have ever been seen around here since flying saucers were invented. They are also seeing new models of them, such as flying tops, orange-colored lozenges, flying mushrooms, and flying cigars, called *les Churchills.* Some of these phenomena are reported to be noisy, others to be as silent as the stars; some are said to fly faster than anything ever seen before, and some—the most alarming—are said to stand still in the night sky. Near Lille, a hundred people living over a broad countryside individually saw crescents dancing arabesques in the sky for twenty-five minutes, and announced the same to the country police. Near Briançon, a flying cigar was seen by a gendarme, a village mayor, and a respected hotelkeeper. West of Deauville, a taxi-driver saw an incandescent red-and-green disc with a phosphorescent tail, and so did some sailors close to shore. In the Nièvre, three citizens in different places saw flying cylinders, orange when they flew horizontally, dazzling white when they proceeded vertically. Not far from Paris—at Coulommiers, where the delicious cheese of that name is made—a member of the Société des Ingénieurs Civils de France, who is also a former aviator and an officer of the Aéro-Club de France, wrote a letter to the local paper anatomizing the object he watched for twenty minutes over his country house. It was something with "a thick circular wing, gyrating around through space with a noise like a jet plane," he wrote. "I am not the victim of hallucination." The second mate of a French cargo ship wrote his company's office in Brest that en route to Antwerp he and two sailors on watch with him saw "a kind of giant red star rising from the southwest, four times the size of any star we have ever seen, and at a point in the sky where no star was due," and set forth the phenomenon's date, its latitude and longitude, and its celestial movements during nearly two hours. The Governor of the French Ivory Coast received a communication from the Administrator of Danané, who declared that for half an hour on the night of September 19th he, his wife, many savages, the chief French doctor, and a missionary priest watched an unidentifiable oval flying machine equipped with a cupola and searchlights, which left two luminous halos behind when it disappeared. The Governor has demanded an investigation. A road mender in the Marne and a baker in Finistère have both met little men from Mars, wearing hairy suits—the baker's visitors also having "eyes the size of crows' eggs." In the last month, more than a hundred reports of strange flying machines seen nearly all over France, but never in Paris, have been printed in the Paris press, with the names of those who saw them, and often with the signed statements given to the local police by observers *"digne de foi,"* or worthy of belief. Among these worthy observers are men like a Conseiller Général des Alpes-Maritimes, a police inspector of the city of Nice, reputable mechanics, educated electricians, a schoolteacher on the Island of Oléron, rich butchers, a wealthy Normandy farmer and his wife, the father of a famous French bicycle racer, and the peasant father of a girl who would not believe what he saw until he dragged her from bed to look and then to the gendarmerie to report. A deputy in Parliament has just asked Mendès-France's State Secretary for Air what it is doing to explain all these phenomena.

November 9

The death in his eighty-fifth year of Monsieur Matisse, as he was always called, brought an end to his long, orderly, and individual reign of genius as *"le roi des Fauves,"* or the king of the artistic wild beasts. He and the young Fauve painters with him in the celebrated 1905 Salon d'Automne—Rouault, Derain, and Vlaminck—along with Braque, who joined them later, founded contemporary French art but no dynasty, because the others soon abandoned Fauvism for painting realms of their own, and although M. Matisse continued to be the emperor of Fauve color, he has left no heir. Early isolated, the revolutionary Matisse, born in Picardy, combined his precise Northern character with the sunny chromatics he discovered in Morocco, and developed into France's greatest modern classic painter. An indefatigable perfectionist, he once made (or so he said) three thousand sketches for one painting. Art became the unfaltering technique of his hand, variety a calculated part of his restricted, disciplined pattern, and only pure color remained free. He was a professorial and magical creator, whose improbable and dazzling Moroccan odalisques lacked all sensuous qualities except the voluptuous coloring he gave them. "If I met a woman on the street who looked like my paintings, I would faint," he once admitted in terse explanation of his theory of style. "I do not paint women, I paint pictures," he also said, summing up the artist's problem of composition. Just the same, his female figures have furnished much of the recognizably human art population of France over the past fifty years. His alternate title to fame came four years ago, when he completed his chapel for the Dominican nuns at Vence, with its modernist black-and-white Stations of the Cross.

M. Matisse was an explicit, logical talker. While painting, he enjoyed listening to gossip, which he did not repeat. He was a born organizer. Two years ago, on hearing that a M. Guillot, municipal councillor and pastry cook in Matisse's birthplace, Le Cateau, was planning a one-room museum of Matisse souvenirs in the town hall, M. Matisse took over. He sent color schemes for the walls, exact measurements for hanging his drawings precisely (one metre and three centimetres from the floor), and also seventy-five original art items, including his 1891 charcoal sketch of a male nude (which earned him a refusal at the Paris Beaux-Arts), some sculptures, some textiles and tapestries he had designed, and two paintings—a collection valuable principally for its intimacy. Though his pictures figure in all important modern collections, especially in the United States, two of the biggest agglomerations of his early works are temporarily invisible to the general public—seventy-five in the Barnes collection, in Pennsylvania, and fifty in Soviet Russia, including the panels "The Dance" and "Music," which he painted for the collection of his White Russian patron, Sergei Stchoukine. With the revolution, Stchoukine's collection was nationalized and put in the Moscow Museum of Modern Western Art, but after the Second World War, the museum's treasures were officially condemned as "artistic bourgeois servility." Their whereabouts was lost in mystery until last summer, when the magazine *Nouveau Femina* printed recent snapshots of the collection, and in particular of the Matisses, showing them unframed, hung by the dozen on portable racks, like washing, and kept locked from the public eye in the attic of the Hermitage Museum, in Leningrad, whose catalogue does not mention Matisse's name. However, the greatest unseen collection is the reported two thousand of his pictures that were owned by M. Matisse himself, many of them stored in a Nice bank. As he once said, *"Ce sont des valeurs."* Their value today is many millions of dollars.

There was difficulty at first in finding a French publisher for General Charles de Gaulle's "Mémoires de Guerre" (of which the opening volume has just appeared), owing to the astronomical price the General demanded for the manuscript and for various rights. There was also a feeling in publishing circles that the memoirs might not sell, because he has lost popularity through his retirement from politics, because his R.P.F. party is splintered, and so on. The highest—and, it now turns out, the luckiest—bidder was the firm of Plon, which is said to have paid him, for only this first volume, a twenty-million-franc advance (about fifty-seven thousand dollars) and, instead of retaining the usual half of the revenues from foreign and serial rights, ceded him forty per cent of the publisher's share. The first regular edition of the first volume, "L'Appel" ("The Call to Arms"), has sold out like lightning. And the limited and numbered de-luxe Holland-paper edition, at twenty-five thousand francs (or more than seventy dollars) a copy, as well as all the other numbered editions on alfa or pure-linen paper, priced at from twenty-one hundred to forty-five hundred francs a copy and reserved for "Les Anciens de la France Libre" and for "Les Membres de Touts les Associations Combattantes et Résistantes de la Guerre 1939-45," were heavily over-subscribed a month before publication. De Gaulle's memorialist style is nothing like Sir Winston's. At its best, his writing has the elegance and precision of fine eighteenth-century French literary style, which is a kind of patriotism in itself. He has a rich vocabulary of simile, recruited into the service of his critical mind by years of cultural reading. The aristocratic detachment of his recollections of his many enemies has been much commented on here, especially this brilliant word portrait of Marshal Pétain: "What a current swept him away, and toward what a fatal destiny! The entire career of this exceptional man consisted of his long effort toward self-repression. Too proud for intrigue, too strong for mediocrity, too ambitious to be froward, he nourished in his solitude a passion for domination, hardened over in time by his consciousness of his own worth, by the setbacks that had waylaid him, and by the contempt he had for others. Formerly, the bittersweets of military glory had been lavished on him. But glory had failed to gratify him fully, because she had not given herself to him alone. Then, suddenly, in the extreme winter of his life, events offered to his gifts and to his pride the opportunity, so long awaited, to burgeon to the full, though on one condition—that he accept disaster as the banner to decorate his fame. . . . But old age betrayed him to the uses of dishonest people, hiding behind his majestic weariness. Old age is a shipwreck. So that nothing might be spared us, the old age of Marshal Pétain was to be identified with the shipwreck of France." Up to the week of publication here, the English-language rights to the memoirs had not been sold, saving some translator a bad time, since the General's French defies easy translation.

Many of Mendès-France's well-wishers wish, for his sake and theirs, that he would stay at home right now, rather than distribute his energies visiting Canada, Washington, and New York. He has been demonstrating in his speeches, activities, weekly fireside radio talks, journeys, and projects on the home front the same talent for programmed momentum that distinguished him in his early, dramatic dealings with vital foreign affairs at Geneva and London, when he molded a new diplomacy for France. He has so much under way that those who believe in the possibility of his renewed France, uprooted from its immobility, feel that he should not drop his hold even for ten days in America, for fear that something may start slipping into reverse behind his back now that his political ill-wishers are recovering their wits and organization. During his five months of power, he has governed in a new kind of political space that he created for himself—above the political parties—and what has upheld him has been public opinion. His popularity with the dominant majority of the French people and with the younger element of Parliament—for he is the leader of the young-generation postwar politicos who are daring to try to do what the old politicos have only talked about—is what has given him the elevation he needed in order to operate in his new political style. But to set in motion the economic, social, and administrative reforms that undoubtedly are his greatest aims, and that Parliament has voted him special powers to tackle, he will have to come closer to earth and will have to rally support in the parties and give pledges somewhere—always the fatal danger of French parliamentarianism on the lower level. Mendès-France is now passing from his first stage, the phenomenal, into his second stage, the actual.

The French say that in addition to having Bidault's M.R.P. party, which has now become Right Wing, and the Communists bitterly opposed to him, Mendès-France has two elements in society against him. First are the all too numerous small manufacturers, whose poor organization, low production, and abnormally high profits make them a losing proposition for the national economy, and whom his reforms would first wipe out and then relocate more usefully. Second are certain high-tariff, power-lobby, and monopoly groups, some of which have been receiving enormous government subsidies; he has already collided with the beet-sugar-alcohol group and a number of big textile interests. He may soon have against him the *bistro* owners and barkeepers, to judge by his strong feelings about the disaster of France's rising alcoholism. A recent graph in *Le Figaro* shows that the French annually consume more than three times as much alcohol as Americans, with all their dry Martinis, do. Mendès-France is the first French Premier ever to drink milk in public, which his wife says he learned to do in the United States in the course of his short stay during the war. His announcement that, beginning with the New Year, France's surplus milk will be distributed to schoolchildren and to the boys in the French Army has caused consternation among French mothers, for the French think that no drink is as bad for the liver as milk. The *Mendésistes,* as his supporters are called, believe that the next three months will tell the story of how far his courage, lucidity, and exceptional intellectual grasp of economic affairs will be permitted to go in making the up-to-date France that everybody has been telling the French they ought to have.

December 8

In its weather, this has been a strange year indeed. It is ending with warmish air in December after a cold June. For Thanksgiving dinner, some Americans living near Paris plucked basketfuls of ripe red raspberries from their walled garden. For Christmas, if the genial late warmth holds, the French living on the Loire will have baskets of roses to put on their table with the

Christmas goose. There have been unusually vicious windstorms —nothing like the American hurricanes named after girls but nevertheless blowing sixty miles an hour. The wheat harvest was larger even than last year's, and bread has just gone up in price. All Western Europe is more prosperous than at any holiday season since the war.

December 23

At the year's end, the Musée du Louvre has received what could rank in art circles as the finest gift of the festive season—five van Gogh paintings, four Cézannes, and one Pissarro. The Cézannes and the Pissarro have never been seen before by the public, and only two of the van Goghs were ever shown— fifty years ago, at the 1905 van Gogh retrospective exhibition here in Paris, when they were little liked. This munificent gift has been bestowed by the elderly recluse Paul Gachet, and it comes, as did two others he made in 1949 and 1951, from the collection of paintings that the painters themselves gave to his father, the notable and eccentric Dr. Paul Ferdinand Gachet, whose inland portrait, painted in his garden at Auvers-sur-Oise and showing him in his fantasy commodore's white cap, was van Gogh's most popular work done in those last fertile weeks just before his suicide. To render justice to that Ile-de-France village, in its day only less famous than Barbizon as an artists' haunt, and to honor both of the Gachets' devotion to the art fostered there, the national Musée de l'Orangerie is holding a show called "Van Gogh et les Peintres d'Auvers-sur-Oise," a touching and, in parts, exciting exhibition of all the Gachet gifts, plus paintings (some lent by American museums and collectors) by all their artist friends who worked or visited there. The simplest of the hitherto unseen van Goghs is a bouquet of "Roses and Anemones," perhaps unfinished, since it lacks any typically anguished striations of color, but an unbeatably beautiful small composition. It was given in gratitude by van Gogh's brother Theo to the then seventeen-year-old Paul Gachet for having kept watch all night by Vincent's bedside as he lay dying. Next in gentle strength is an exquisite, creamy painting of "Mlle. Gachet in the Garden," blooming with white roses. The biggest and most violent picture, still vibrating with van Gogh's insane Midi style, is "Dr.

Gachet's Garden," the first painting van Gogh did on the family premises, showing a giant yucca plant and cypresses against the background of a fiery-red rooftop. These last two pictures have been kept so private since the summer of 1890, when they were painted, that they had never even been photographed until plates were made for the Orangerie catalogue; art experts had known of them only by Vincent's letter of June 4th to Theo, in which he wrote, "I painted two studies at the Doctor's, which I gave him last week," and then described the cypresses and rose vines *"et une figure blanche là-de-dans."* The two other new van Goghs, both exhibited in 1905, are "Thatched Houses at Cordeville," a grand example of his ascending, hillside equilibrium in design, and "Two Little Girls," a serious-faced version of those two ugly, small peasants who are already famous in their smiling version. Among the four hitherto unseen Cézannes, all painted at Auvers in 1873, his backward-looking, so-called "baroque," manner is visible in "Bouquet of Yellow Dahlias" and, especially, in the fascinating jumble of table objects—including a medallion, an inkpot, a roll of paper tied with ribbon, and some of the inevitable fruit—which old Gachet humorously entitled "Cézanne's Accessories," the name the painting is now known by. The two other new Cézannes show his future style, in the boldly geometric houses of "A Corner on Rue Rémy in Auvers" and in "Green Apples," a small, truly great painting. Among the other painters in the Orangerie show—all of them the Doctor's friends—are Daubigny, Corot, and Daumier, as well as Pissarro, whose hitherto unseen Gachet canvas is a snow scene called "Chest-nut Trees at Louveciennes."

The three Gachets were peculiarly lucky. None of the Auvers painters (all then still struggling either for fame or money) meant to give the garrulous, pushing, art-loving Doctor any important paintings—except van Gogh, who always shed his masterpieces wherever he was, like a fruit tree dropping ripe fruit. The unique charm of the Gachet collection today is its homely, connecting intimacy—its pictures by painters painting each other or the Auvers houses or the Gachets' house or the Gachets themselves or the Gachets' endless bric-a-brac, so useful for still-lifes, such as the Doctor's tasteless gray Japanese vase. It was used by both Cézanne, in "Bouquet of Yellow Dahlias," and van Gogh, in "Roses and Anem-ones," and, like the Doctor's white commodore's cap, it is on view today in an Orangerie showcase. The Gachet collection cost the

Gachets nothing except the Sunday dinners they offered to their artist friends. But in donating their twenty-nine pictures to the Louvre, son Paul and—until she recently died—daughter Marguerite, of the white rose garden, were generously giving away a large fortune. Though Cézanne was careful not to offer the Doctor his Auvers masterpiece "The Hanged Man's House," he did hand over to him seven paintings inescapably marked by genius and ultimate value. And in the total of eight Gachet van Goghs now the property of the French state there are three almost priceless masterpieces—the Auvers church, the pensive portrait of the Doctor in his seafaring cap, and Vincent's last self-portrait, with his tormented face surrounded by what look like blue flames. Paul Gachet, now eighty-two, has apparently come to the bottom of the family bin and has nothing left to give away. Austere, black-browed, looking like a gaunt Daumier drawing (the Doctor was red-headed, roly-poly, and over-sociable), he is frugally finishing his days in the emptied Auvers house.

1955

January 5

Twice in four months the French Parliament rejected the idea of West German rearmament—once at the end of August, when it voted down the European Defense Community, and once at dawn of the day before Christmas. And then (one can hardly say suddenly after four years of shilly-shally), two days before New Year's, Parliament voted for it and for a Bonn army of half a million men. The changing vote did not reflect any change of mind. Because of history and memory, almost to a man the people of France, like their deputies, have been—and still are—opposed to giving any German any kind of gun. Premier Mendès-France was indeed accurate when, in the course of persuading Parliament to ballot his way, he said that it was "despite the repugnance of the French people" that French participation in West German rearmament had to be ratified, "as the sole means of preserving the Western alliance." He also told Parliament that Washington had come "within a millimetre" of organizing this rearmament of Germany on its own while the deputies were still needlessly arguing—many of them with intelligence and French logic—as if they could prevent it.

When Assembly President André Le Troquer, in the evening clothes that are his garb of office day and night, read the figures of the final vote—two hundred and eighty-seven for German rearmament, and two hundred and sixty against—there was a long moment of expressionless silence. Nobody applauded, nobody moved, no deputies inclined their heads together for a quick private comment; there was not the sound of a whisper. There was nothing but silence, maintained for a suspended instant in the presence of a vital act of history. Then Le Troquer quietly added that the session was closed, and the Communists, on their left benches, safely raised their first protesting shouts. Excessively disorderly acts or comments while Parliament is still officially sitting can result in the offending deputies' being fined as much as a day's pay, but now the Communists could vent their fury freely. They jumped to their feet as a group, the male deputies shouting and the female deputies screaming in a jumbled clamor, from which at first only single words came clear—"Assassins," "Traitors," "Shame." Then one woman shrilled a whole audible phrase—"Ten years ago, French women were being tortured to death at Ravensbrück!"—and a man behind her bellowed, "Mendès, you'll have a chance now to know what German crematories are like!" The Premier's pale, calm Jewish face did not alter; it remained masked, attentive, motionless. The uniformed ushers now rushed into the public galleries, calling loudly, *"Evacuez la salle! Sortez d'ici!,"* and even pulled some startled visitors from their seats, so that they might not witness any further disorders. When news of the ratification spread to the Quai d'Orsay sidewalk, in the early evening darkness, the policed and barricaded lines of Communist delegations and sympathizers began singing "La Marseillaise," as if a Russian had written it. As they had done during the preceding ten days of debate, many men and women in the propaganda line held up banners inscribed with slogans like "The French Resistance Veterans Say No to German Rearmament" and "Don't Forget Auschwitz." Some displayed enlarged photographs of piled-up bodies at Buchenwald, of dead French faces bloodied by Gestapo tortures during the Occupation, or of emaciated, shaven-headed concentration-camp survivors, variously identified by those holding the photos as "My brother," "Papa," and, occasionally, "Me." For once, Communist propaganda merely coincided with French public opinion; its slogans and pictures only revived French memories of what the Nazi Germans had done, and solidified the angry fear of the French that such a race would do the same thing again if it were given guns.

On the other hand, what the Russian Communists have done habitually to the map of the world in their postwar program of occupations was, strangely, little mentioned in the ten days of debate, even by the five hundred or more anti-Communist deputies—an odd omission, considering the increasing Vietminh encroachment right now on the so-called democratic Vietnamese, whom France fought for in Indo-China for seven years. There was one exception—the militant speech by Socialist Deputy Guy Mollet, almost the only deputy who truly favored German arms. "We have to have them," he declared loudly, "because since the war Russia has kept millions of men under arms, because in the last ten years Russia has destroyed the liberties of so many people, and because all the troubles today come from Russian expansionism." (For this attack, the Communists denounced him as "a dog of a Socialist" during their post-ballot uproar.) In the minds of many American observers here, Washington officials typically gabbled out of turn once more when, in the midst of Mendès-France's desperate efforts to persuade the deputies to accept the American view of the urgent necessity for bringing in Germans, and thus swelling the ranks of the soldiery defending Western Europe, Washington suddenly made the bland announcement that the United States was reducing its own army, since the use of atomic weapons could substitute in a defensive action for great numbers of men.

It is a matter of history that the rearmament of West Germany—it was to have been accomplished with ten divisions of Germans—was first proposed, in 1950, by Secretary of State Dean Acheson, alarmed by North Korean aggression. The French countered with what became known as E.D.C., a European army that would have had some Germans in it, and even with vague supranational notions of forming a United States of Europe. However, although the French had created E.D.C., they were still alarmed at the thought of giving any arms to any Germans under any circumstances, and eventually voted against it last summer. Now, for the new year of 1955, they have resignedly ratified just about what Acheson outlined in the first place, except that the new agreement gives the Germans twelve divisions and provides sovereignty for Bonn. Thus the French lost what might have been their chance to form the mold for modern European history, and, moreover, were forced to take on the familiar old setup that they had most feared from the start. On December 31st, several Paris newspapers ran immediately beneath their final report on the Assembly's ratification the news that sixteen big German firms of the Ruhr, where the German armament crowd has always had its headquarters, are ready to start West Germany's first atomic pile, at Düsseldorf, just as soon as the Paris treaty starts functioning.

January 19

The Seine is in flood. The *quais* are under water, so that the river looks almost half again as wide as it is, and the bridges are black-edged with citizens hanging over the parapets to stare their fill at the dramatic swirling, yellowish stream beneath, the favorite bridge being the Pont de l'Alma, where the water is already halfway up the thighs of the famous Zouave statue between the arches. It came to his neck and almost drowned him, as the saying goes, in the great floods of 1910, when rowboats replaced carriages on the Place de la Concorde, and it is now climbing up him at the rate of about an inch an hour. All traffic on the Seine has been stopped, and boats have been ordered to tie up and register their positions with the Prefecture of Police, in case some have to be called to evacuate riverside dwellers in low-lying Paris neighborhoods. All the rivers in France are alarmingly high. The Marne, whose rain- and snow-swollen headwaters take eight days to flow into the Seine, is already above its 1910 high mark, and next Monday it and the Yonne, whose water, even now overflowing, takes only four days to reach the Seine, are together expected to make an accumulated high tide in Paris that will be six metres, or about nineteen feet, above normal level. The Paris skies over the waters are clear and sunny. The recent wind and rain storms at sea and on Channel lands were so fierce that numbers of sea gulls were blown into town; some now follow around on foot after the Louvre pigeons, hoping to pick up some of the crusts occasionally tossed their way by children, and even fly down into Parisians' back yards and gardens in the hope of charity. This has been an abnormally tempestuous new year, blowing first hot and then cold in no reasonable rhythm.

February 2

The floods in the cities and the countryside are now subsiding. A lot of France's arable land was under water. The winter wheat and oats are largely drowned. Paris itself was mostly spared. On the Ile de la Cité, there was some awkward flooding of the underground cells in the Palais de Justice, where prisoners are kept until they are brought upstairs for trial. At the Institut, thousands of books were rescued from seepage in the basement stacks; fortunately, the famous Mazarine Library was not affected. Along the Quai de Passy, anchored boats rode higher than the parapeted street level, and from there to the west nothing could save the low-lying suburbs and towns, where, for nearly thirty miles—from the motorcar factories at Billancourt, through Le Pecq, and on to Poissy and Mantes—the risen river simply flowed into people's lives via their shops, parlors, bedrooms, and factories. In the towns on the periphery of Paris, two hundred factories were closed, and to the fifty thousand workers the government has now promised out-of-work pay amounting to seventy-five per cent of the hourly minimum wage for single workers and eighty-five per cent for family men. In and around Paris, the organization of an estimated seventy-five hundred professionals and volunteers for rescue work and flood fighting was astonishingly efficient. On the job were six companies of French Army engineers; Paris police; highway police; every fireman and ladder within reach; municipal masons and ditchdiggers; L'Union des Femmes Françaises, which persuaded people in dry homes to take in the castaways; La Fédération Française de Canoë-Kayak, which operated its fragile pleasure craft for delivering medicines to the sick; Boy Scouts; local American Army units, which gave raincoats and invaluable rubber boots; the Pope, who gave a million francs to the refugees; and the city of Warsaw, which topped him with ten million. Thousands of student volunteers sandbagged river parapets, working under the scholarly and apropos group title of Fluctuat, taken from the old Paris coat-of-arms motto, *"Fluctuat nec Mergitur."* Others worked in a group called Zouave, in honor of the Pont de l'Alma statue that always serves as a floodmark and on whom the river this time rose to the elbow. In the present unseemly spring weather, France is cleaning up her mud and trying to dry out.

Floods were also unkind to Pierre Poujade, the bookseller from the little town of St.-Céré, in the Lot, who is the leader and national stump speaker of a nineteen-month-old revolt of more than half a million shopkeepers and artisans against what they claim are the iniquitously high and unjust taxes the government imposes on them. He advertised that two hundred thousand delegates of his Union de Défense des Commerçants et Artisans would meet in Paris last week to air their grievances, but the inundations made transport irregular, the state railroads refused to issue the cut-rate tickets customary for pilgrims or mass delegations, and the Department of Bridges and Roads refused to permit fleets of hired buses on the waterlogged highways. The fifty thousand delegates who eventually turned up in the Parc des Expositions at the Porte de Versailles looked to be unusually pleasant, uncynical, modest French. There was a Paris rumor afterward that Poujade had ranted like a coming dictator while his men followers roared and his women followers screamed, threw kisses to him, and swooned with emotion. Actually, the women at the meeting seemed to be mostly married shopkeepers used to keeping their eye on the money till; as Balzac made clear, *madame la caissière* is not given to fainting spells. Poujade himself, aged thirty-four, is a chubby, handsome, curly-haired brunet and an earnest and perspiring talker, whose statistics are usually wrong but who makes sense to his followers because they agree with him. In his speech, he claimed that recent government tax reductions merely martyrized the little fellow and favored big business, and that "Mendès-Lolo" (French baby talk for milk-drinking Mendès-France, whose name was booed, probably by the many café owners who are Poujade followers) wants to wipe out small tradesmen and artisans—"though for centuries we have been the backbone of France"—in favor of mass production and massive sales. He said that those of his followers who had chased nosy tax collectors out of their shops when they came to inspect the ledgers were only fighting for their liberties. "We do not want the moon," he said, and merely asked an annual tax exemption for *les petits commerçants* on their first three hundred and sixty thousand francs (about a thousand dollars); amnesty for all delinquents who have committed tax frauds; and a strike by his followers against paying any further taxes (these last two projects were enthusiastically cheered) until tax-code reforms should be established. Government tax collectors say that fiscal frauds among such small businesses are indeed widespread; that, at latest reports, four-fifths of the nation's tradesmen—including butchers, grocerymen, plumbers, masons, and *bistro* keepers, the groups that unquestionably skim off the cream—declared monthly earnings equivalent to eighty-nine dollars, only twelve dollars more than a factory worker's wage. However, financial experts say that Poujade's proposed tax revisions have some justification. In any case, his organization's threat that it can influence coming elections is already worrying some deputies. Meanwhile, the authorities have asked the Directeurs Généraux des Impôts et de la Comptabilité Publique, two high officials in the Finance Ministry, to prepare a report on whether they can sue Poujade for "inciting the public to refuse or delay payment of taxes."

February 15

Right now, the bourgeois politicians here are strain‧ ing their capacities in an effort to scratch together a government to replace that of Mendès-France, which they overthrew. During that violent, deafening half hour of shouting and banging desk tops at four in the morning, when a majority of the deputies tried to prevent Mendès-France, just voted out of office, from speaking from the tribune, one of the oddest figures on the scene was the Assembly's Secretary-General, the only official who was perfectly calm. A familiar slender silhouette on the rostrum, with a distinguished shock of blue-gray hair, he is known to few Assembly visitors by his name, which is Emile Blamont. His duties demand discretion and a strong spine. He is the nonpolitical permanent authority on parlia- mentary procedure. During decisive debates, he sits with his back to the deputies, uncomfortably perched on a red velvet postilion seat that is attached to the rostrum beside the chair of the Assembly President, or speaker, so that he can lean down to talk confidentially into that official's ear if he is consulted or if he thinks that procedure or questions of order are being misinterpreted. In his replies or advice, Blamont must be quick-witted, explicit, and accurate, with an infallible memory for parliamentary and constitutional affairs, and must by his demeanor display great tact (as he did during the hubbub over Mendès-France), since he does not represent the authority of the people, as the deputies do—a responsibility they are very jealous of—but only the authority of the Civil Service, which has appointed him as its functionary, making him a nonpolitical and constant outsider. In his administrative capacity, he has all the other parliamentary nonpolitical functionaries under him, such as the relays of stenographers who take down the debates, and the silver- chained *huissiers* in their long tailcoats. After passing his civil-service examination in 1930, Blamont entered the Assembly organization in a junior role, was in London during the war until General de Gaulle founded his Algiers Assembly, and then became its Secretary-General, there being plenty of new politicians down in North Africa but he being the only functionary who represented continuity with the Paris Parliament. After the Liberation, he took up his present post. He has great admiration for the British House of Commons, and knows his opposite number there, Sir Edward Fellowes, who certainly has never had to remain impersonally calm in the face of any such hurly-burly as attended Mendès-France's fall.

March 2

"The prophetic oak has been brought low. He was one of the grandeurs of our time." "After Valéry, after Gide, now, with Claudel, disappears the last of the *grands maîtres* from the end of the nineteenth century." With such threnodies was the passing of the poet and dramatist Paul Claudel, aged eighty-six, recently mourned by Paris newspapers. They carried their appreciation of Claudel's literary gifts, and their expressions of national loss, on their front pages, along with the announcement of the Ministerial makeup of the new government. Claudel's phenomenal gifts made him an awe-inspiring figure. He was a noted dramatist; a profound mystic and religious poet; an almost lifelong diplomat, having been Ambassador to Tokyo from 1921 to 1926, during the earthquake, and to Washington from 1927 to 1933, during the depression; a passionate revivalist of the Roman Catholic faith; an authority on Franco-American finances, having served with the French Economic Mission at the end of the First World War; author of some half-hundred volumes, which he said he wrote in daily half-hour stretches; and a director of the rich Gnome & Rhône, France's big airplane-engine factory.

The general public knew Claudel only through his theatre pieces, which have lately been fashionably popular, and thus impressive to most people, if, in accordance with the inevitable dual reaction to him, very tedious to some, who found his "Le Soulier de Satin" merely the longest play they ever sat through. "L'Annonce Faite à Marie," a medieval mystery play, written in 1912, that was given by the New York Theatre Guild as "The Tidings Brought to Mary," now functions as a French classic in the state theatre. It has just been given a revival at the Comédie-Française, badly acted but with a glorious gala opening, which Claudel attended. After the opening, the powerful old Christian, with his own words about the glory of God still fresh in his ears, suddenly died, on Ash Wednesday.

Some Parisians have remarked that the nineteen days when they were without a government created a dangerous euphoria. France's various grave problems—North Africa, a new tax scale, a reformed constitution—were temporarily forgotten, because nobody in Parliament was talking about them, or even talking about doing something about them. The French citizens, thrown back on their own devices, lived and functioned as usual, except that they developed a vague feeling of national solidarity, of being the people, as against those six hundred and twenty-seven deputies who had been elected to serve them and, as a sign of their customary quarrelsome inefficiency, had left forty-three million citizens in the lurch, and thereby convinced of their own superiority.

Pierre Mendès-France is now called the Phantom in the Corridor, the ghost who haunts the Parliamentary halls with the memory of his regime. It is not easy to see how his successor, Premier Edgar Faure, will be able to settle the important problems that now, after the delay, face the country more immediately than ever. For Faure, holding his present Assembly majority together is like balancing a pyramid of eggs. If he so much as makes a move to do anything positive, or if one of his supporting Ministers so much as lifts a disapproving finger, the eggs, safe only while immobile, can well come breaking to bits around his head.

March 17

In a press conference held late yesterday afternoon, Premier Edgar Faure wryly said that if he and his government fell Friday night, the fall would hit him hard, because it would occur on his twenty-sixth day in office, whereas when he was Premier in 1952, he managed to stay in power forty days. If Faure falls this weekend, it will be because of Pierre Poujade, the pugnacious thirty-four-year-old provincial shopkeeper from St.-Céré, in the Lot, who slammed the door of his little stationery shop in a tax collector's face in July, 1953, and last year founded his Union de Défense des Commerçants et Artisans, an organization, designed to thwart tax collectors, that

Paris had never even heard of six weeks ago. The U.D.C.A.'s expansion in the past six weeks has been a social phenomenon. In January, at Poujade's first Paris meeting, which was made up largely of his devoted provincials and snubbed by Paris shopkeepers, he claimed four hundred thousand followers; this week, he claims eight hundred thousand. His organization is now nationwide, except for a few *départements* in the northeast, which is Gaullist, and a couple in the southwest, which is Basque. February 15th is the date when the French should pay the first third of their year's taxes. In eighteen *départements* of central France—the region where Poujade started, and where he is still strongest—thirty to eighty per cent of the little businessmen and artisans failed to pay their February taxes and closed their modest savings accounts, so that government authorities could not block their funds until that tax money starts rolling in.

This week, Poujade turned to direct action and political pressure in the capital of France. First, he sent Président du Conseil Faure an open letter—described as threatening—that accused him of having caused his Assembly majority to break faith with the U.D.C.A. on rapid tax reforms, which were put to one side in last week's preliminary debate on finances. "The responsibility for the rupture is yours, and you will bear the consequences," he warned Faure. In the same mail, he sent three hundred deputies a letter listing U.D.C.A.'s demands. In it, he said he wants amnesty for tax frauds, which he considers the consequences of injustice; suppression of the tax collectors' right to snoop in taxpayers' account books; abrogation of the Ulver-Dorey law of last summer, which provides harsh punishment for anybody who discourages tax collection; and the appointment of a Parliamentary commission to study tax reforms—with the U.D.C.A. itself, no less.

Poujade bypasses the Ulver-Dorey law, which imposes fines of up to a half-million francs, as well as prison terms, upon people who incite the public to refuse or delay its tax payments, by merely saying humorously to his followers, "Turn off the faucet. Don't turn it on again till victory is ours." There is no French law against tap-turning. He ridicules what is called the *polyvalent* system of tax collection by nicknaming all tax collectors *polyvoleurs* (multiple thieves). In central France, the U.D.C.A. is highly organized, with the result that when a tax collector makes a date with a U.D.C.A. shopkeeper to inspect his books, the shopkeeper calls two U.D.C.A. members, who each call two more, and so on, and within half an hour they have collected a crowd, which either keeps the taxman from getting near the shop door or chases him out of town. This system has been harder to work under the noses of the hardboiled Paris police, but a fortnight ago the newly organized U.D.C.A. branch in Paris temporarily broke up the bidding at the Hôtel Drouot's state auction rooms on some wretched bankrupt shopkeepers' worldly goods. To everyone's surprise, the marketmen of Paris, the fourteen prosperous syndicates of food sellers, and even the rich Butchers' Association have climbed aboard the Poujade bandwagon in the poor man's tax revolt.

Yesterday, Poujade's direct action brought in a small tidal wave of results. Combined-political-party groups like the Independents, the Independent Peasants, and the A.R.S., an ex-Gaullist splinter faction, sent delegations to Faure demanding "profound tax reforms" if they were to back his Friday demand for a vote of confidence. Yesterday, too, some deputies spoke up in Parliament about the pickle they were in, caught between the need for defending him and the risk of offending their aroused U.D.C.A. voters. And the Federation of Finance Officials, which includes tax collectors, all of them furious, held a wonderful press conference, fulminating against Poujade and declaring that some shopkeepers had threatened to nab tax collectors as hostages, had refused to sell goods to tax collectors' wives, and had instructed their own children to put tax collectors' children in Coventry, neither talking to them nor playing with them at recess. "We can no longer tolerate this situation!" the Federation's president cried. "If the government persists in refusing us police protection, we will set up our own defense and launch reprisals. We will refuse to buy from shopkeepers who don't pay their taxes," he added weakly. In addition, yesterday the Parliament's Finance Commission suddenly voted—thirty-five to seven—to make abrogation of the Ulver-Dorey law the price of its support of Faure's special powers on Friday. Last night, *Le Monde,* which spoke with contempt of the Finance Commission's cowardice, carried on its political page seven articles and news items that dealt with *l'affaire Poujade.* This morning, *Combat* ran five front-page stories on it and an editorial entitled *"Un Précédent Dangereux,"* in which it said sadly, "Poujade has proved the merits of direct action against a decadent democracy."

Whether Faure falls or gives in and stays in so-called power, if Poujade has anything like the number of followers he claims, right now he controls the biggest single-idea social-reform group in France.

March 30

Considering the constant political hot water most French politicians live in, it is a relief to know that a few of them occasionally take time off to practice an avocation. It was not admitted until recently that Premier Edgar Faure has written three detective stories, under the nom de plume of Edgar Sanday. San day is a kind of phonetic joke, meaning *"sans d"* and referring to the fact that his first name is spelled the English way, instead of being spelled "Edgard," *à la francaise.* "M. Langlois N'Est Pas Toujours Egal à Lui-même" ("M. Langlois Is Not Always at His Best"), which is Sanday's last detective story, was printed in 1950 and is still selling. (His first two, written as pastimes during the Occupation, when the Germans were hunting for him, are out of print, and little is generally known of them now except their titles—"Pour Rencontrer M. Marshes" and "L'Installation du Président Fitz-Mole.") The Langlois yarn starts as a highly entertaining suspense story of character, focussed on a mysterious, dull foreigner of that name who suddenly arrives in an elegant provincial town and undertakes the odd business of settling money quarrels out of court, and the odder use of drugged tobacco to paralyze the resistance of his clients, the hapless local rich. After bankruptcies and subtle murders, and some shenanigans by his accomplice, an exotic actress, Sanday weakly lets the pair of them disappear in an equivocal finale. It seems that his family, to whom he read his story by chapters, were indignant at this evasive ending. Premier Faure is a habitual reader of other people's detective stories, especially when travelling, his last favorite being everybody's—Albert Simonin's classic in argot, "Touchez Pas au Grisbi."

April 13

You can tell it is spring here mainly because the tourists have burst into bloom. Nature herself is belated. While a million and a half Parisians headed out of town for Easter, half a million visitors, the foreigners among them mostly English, German, Danish, or American, flowed in for the chilly fête as part of the largest weekend holiday migration that West Europe has yet recorded, with citizens travelling in all directions—an inspiring parade of prosperous pocketbooks.

Spring being the proper season for major art, "Salut à la France, Hommage Culturel Américain," the biggest presentation of American culture ever brought here, has opened its first exhibit, "Cinquante Ans d'Art aux Etats-Unis," at the Musée d'Art Moderne. It is no secret that this salute is an expensive propaganda project, launched in the hope that it will counteract part of the ill will that some of the Americans over here and some of Washington's foreign policies have gained for us since the war. It is also no secret that the whole scheme was viewed as dubious by certain State Department and Congressional figures.

Doubtless to everyone's surprise, this modern-art show has attracted some of the largest crowds the Musée has ever had, and has received mostly thoughtful critiques, often praising the very artists the same critic condemned two years ago. New York's Museum of Modern Art, which assembled the show from its own collection of paintings and sculptures, has put on display a group of photographs, some advertisements, samples of typography, and blown-up pictures of such examples of American architecture as Frank Lloyd Wright's glass tower in Racine and the Chicago glass apartment house of Mies van der Rohe, which the French consider miraculous. The entire Parisian press has given the show attention. The artists most frequently cited have been the elders—Lyonel Feininger and Maurice Prendergast, who were previously scarcely known here, and were immediately traced to Cézanne and Bonnard and rated as important. Everyone, rightly or wrongly, has been identified with a school: Marsden Hartley and Arthur Dove with the Blaue Reiter; Jackson Pollock and Arshile Gorky with Expressionism; Morris Graves and Hedda Sterne with the romanticists; Mark Rothko with the constructivists; Mark Tobey with the Orientalist Intra-Subjectivists; and the Ecole de la Poubelle, or ashcan painters, with the Zola realists. John Marin was unanimously selected as the only painters' painter. In general, the critics said that American figurative painting is socially ironic, never imaginative, and that American abstract works tend to be young and dry. Some of the criticisms were unconsciously funny, such as that in *L'Express,* which kindly explained American artists by saying, "Cut off from Europe in 1940 by the war, or in contact only with European émigrés, the young Americans had to follow their own path almost without guidance." All the French critics showed an almost morbid interest in pointing out how many of the artists have Continental names, but only the weekly *Carrefour's* critic went so far as to say, "I perceived in the exposition no American art, no American style. . . . Nothing but foreigners."

Parliament is on vacation, so that it can keep a finger on the country's pulse during this week's cantonal elections, but the deputies' absence does not mean that political life in Paris has ceased. The government is straining to settle its knotty North African colonial-reform questions before April 23rd, the start of the great Moslem festival of Ramadan, which is always a period of high tension, owing to the people's prolonged fasting and then the final, explosive celebrations and feasts. In the meantime, the native leaders of the Comité Maghrebien, representing the Istiqlal, the Neo-Destour, and the Liberation Front of Algeria, have all gone off to the conference at Bandung, in Indonesia—until recently a Dutch settlement, the French keep reminding each other while wondering about the future of their own colonial possessions—to meet other non-Europeans, including sixteen prime ministers, among them Chou En-lai, Ngo Dinh Diem, Nasser, and Nehru.

The destructive squabbling between France and America over the Indo-China imbroglio has made for renewed worry about French and American foreign policies while the Communists seep farther southward into Vietnam, giving French parents who vainly lost their sons there a doubly sad anniversary, for the fall of Dienbienphu occurred almost a year ago.

April 27

The long shadow cast by the Bandung meeting of Afro-Asiatic premiers and delegates of color, speaking their minds for the first time in modern history without the white man's domination while they deliberated as racial authorities on colonialism (which they condemned), may possibly have aided the French government this last week, since there were Tunisian delegates at the conference, criticizing the do-nothingness of the French in settling the Franco-Tunisian crisis. With the eyes of the world thus focussed on colonial problems, the Faure government was able to reach an accord with the Protectorate of Tunisia, by which, for the first time, France gave the Tunisian natives limited home rule. It was not too soon, after thirty years of reform talk and broken French promises, punctuated lately by the Tunisian natives' murdering of white French colonials as a sign of their impatience. Although the protocol was signed by Edgar Faure, as the French Premier, and Tahar Ben Ammar, as the Tunisian Premier, Faure was signing, in a way, for Mendès-France, who, as Premier, last July solemnly engaged France to grant this long-promised autonomy, and the Tunisian official, in another way, was signing for Habib Bourguiba, chief of the Neo-Destour Independence Party, whose consultation with Faure two days before established him as the real intermediary in arranging this new Franco-Tunisian step forward. Bourguiba's suddenly admitted importance in Paris was an extraordinary story in itself, even for this cultivated rebel against French rule, who, through his education at the Sorbonne, is more familiar today with the libertarian Voltaire than with the Koran. Arrested three years ago for his activities by the Resident General of the Pleven government, he was later exiled from the Tunisian mainland to some unnamed island off the coast, which his son, in his turn a student at the Sorbonne, identified to friends here as the island of La Galite, adding that it was uninhabited except for his father and the three soldiers guarding him, and that his father had been forbidden to take books or any other reading matter with him, thus being exiled even from the companionship of print. Removed from his island in one of Mendès-France's first steps toward relieving dangerous Tunisian tension, he was brought to France and settled comfortably outside Paris in *résidence surveillée,* which roused against Mendès-France the fury of Tunisia's politically powerful, colonial conservatives. These white diehards were affronted anew last week at seeing the native they most resent consulting freely with the head of the French government on this matter of limited home rule, which liberal colonials and most Paris politicians realize is the minimum that could be belatedly offered to the Protectorate in 1955.

May 11

Historically, this has seemed to most of the French the strangest week since the end of the war, because all that was postwar ended, too; Western Germany is now not only sovereign but an ally. It was also the most important week since the end of the war for anniversaries, memories, and ironies; for hopes and plans for one more new Europe; for vitally important international diplomatic talks and sharp disagreements. The highest British, American, and French figures in foreign affairs have had their first meetings here in their new relation with Western Germany as friend and equal, and if, of the original Allied quartet, Russia was absent, she was certainly present in the conferees' thoughts one way or another. It must have been due to the haste with which history is fabricated today by supersonic flying diplomats that although Saturday, May 7th, was the tenth anniversary of Nazi Germany's unconditional surrender, Monday, May 9th, was the day chosen to welcome federal Germany as a sovereign nation, fit to be rearmed, into the bosom of the Atlantic Pact ministerial meeting at the Palais de Chaillot. The French were not the only ones in Paris to have their memories harried by this indelicate lumping together of contrary feelings and events into a bargain weekend in history. The Norwegian NATO delegate, who, like the fourteen other delegates, spoke optimistic words of welcome to the incoming fifteenth power, represented by Chancellor Adenauer, had the greatest reason of all present to hope the moment was indeed "a decisive turning point toward European peace," since he spent two years in the Sachsenhausen concentration camp.

Some French papers stoically printed photos of the Kaiser's old red, gold, and black flag hoisted in the Allied flag circle at SHAPE, but none printed a photo of West Germany's military representative there, General Hans Speidel, who was tactfully dressed in civilian clothes. Only the Communist paper *Humanité,* pursuing the Soviet line against German rearmament even after the event, ran any information about him. In a column entitled "Qui Est Speidel?" it stated that as Rommel's chief of staff, commanding the German forces north of the Loire in the spring of 1944, he, like Rommel, had been intelligent enough to realize that Hitler was headed for disaster, and that he and Rommel had received at their headquarters at La Roche-Guyon the well-known German writer Captain Ernst Junger, who was already preparing a peace treaty for submission to the Western Allies. According to its paragraph headings, this treaty, later published here in book form under the title "L'Appel," was based on the idea of "Europe unified on the base of Christianity and German predominance, destined to avert Bolshevism."

Germany as an entirety is the European hinge on which either the Eastern Russian Communist influence or the Western democratic and American influence must eventually swing, it has been thought here. Adenauer's request for Allied aid in unifying the two Germanys, and Russia's proposal at the London disarmament conference for the suppression of atomic arms and the withdrawal of all Big Four troops from the two Germanys, were like psychological and diplomatic bombs in themselves as they fell on Paris. There will be plenty to talk about at the coming Big Four conference, which looked so unlikely a month ago and which it is taken for granted here was pushed by President Eisenhower, rather than by his Secretary of State. The French now feel that at least one good thing has come from the rearmament of West Germany by the Paris accords, and that is a concentration of full attention on the ultimate European problems. It is worth repeating that this past week in Paris has been the most important week since the war ended in that memorable month of May ten years ago, when there existed so much simple hope and confidence in what was to come.

May 25

The musical contribution to the American propaganda program called "Salute to France" has consisted of three concerts by Eugene Ormandy and the Philadelphia Orchestra, which made a brilliant impression on Paris music critics. That is to say, its violins were appreciated as wonderfully silky; "I wager there are some Stradivariuses, Guarneriuses, and Amatis among them," one critic bet in print. It was praised for its youthful ardor and disciplined virtuosity but criticized for lacking nuance and for playing almost the whole time as loud as all outdoors. The programs, which it is rumored were imposed on Ormandy by an influential impresario who guaranteed the State Department that he knew what the French liked, supplied nearly everything sure not to please them, except one Beethoven symphony. The French, who still care relatively little for Brahms and less for Tchaikovsky, got both, as well as arias from Handel's "Samson" and Verdi's "Otello," a Bloch symphony, and such French standbys—of which the French are weary—as Debussy's "La Mer" and Ravel's "Daphnis et Chloé," and "La Valse." They also got Brailowsky, at the last concert, in impressive Chopin, but to the critics these were *programmes déséquilibrés,* without enough *vraie musique*—meaning serious classics—and, of course, with nearly no pianissimi. Thus, this part of the Salute to France, for which so much praise was expected, backfired. And a great pity.

June 15

The Peking Opera's fine clowning and acrobatics, the last pleasures on earth Paris was expecting from the mysterious Communist East, proved to be such a hit in the Festival International d'Art Dramatique, at the Théâtre Sarah Bernhardt, that the troupe has been scheduled for additional performances at the Palais de Chaillot. Practically everything connected with the Chinese show was contrary to what had been anticipated. The performers wore no splendid old robes, weighty with hand embroidery, but colorful, gaudy costumes of factory shoddy that looked as if it had never seen a silkworm—easy for the actors to turn somersaults in, as a means of crossing the stage between speeches, in the comic legend-plays that are part of the Opera's surprising repertory. Because of Parisians' familiarity nowadays with dodecaphonic and "concrete" music, even the old pentatonic-scale melodies in the Chinese orchestra sounded quite comprehensible here. Only the high-falsetto operatic singing of both soprano and tenor in Act VIII of the Ming Dynasty opus "The White Serpent" (no other acts from it, or from any other opera, were given, perhaps through instinctive tact) was still painful to Western ears—unfortunately, since this was probably the summit of the company's elegant Chinese classicism of unreality. The lowest point was a dainty Sino-Soviet work ballet featuring maiden tea pickers, pretty as pictures with their rouged eyelids, their smiles, their fans, and their exquisitely idle-looking hands.

The most highly refined clowning was performed by Chang Chun-hua, the Peking Opera's most celebrated acrobatic comedian. His role was that of an innkeeper hunting for an enemy wayfarer in a supposedly unlighted bedroom. The deft game of hide-and-seek, which lasted nearly an hour—the Chinese are more patient than we are, and all their acts were very long—actually took place under bright theatre lights, so that nobody could miss the fun as the two men, silently somersaulting, leaping, and groping, worked side by side as a team, perfectly missing each other. The joke was seven hundred years old, being taken from a Sung Dynasty play, and certainly seemed the ancestor of many of our circus-clown gags. Both actors were clearly in their thirties and were expert gymnasts—fifteen years of physical training reportedly being part of the Peking troupe's apprenticeship, even for falsetto singers. The big production number, with chorus, was "Troubles in the Heavenly Realm," starring Wang Ming-choun, the Opera's greatest satiric mime, as king of the monkeys in revolt against the stupid gods—a role he played like a Chinese Till Eulenspiegel, with the addition of fabulous tumbling and brilliant fencing. "The Yentang Shan Fortress" was a modernized version of a medieval battle scene, with the opposing troops fighting in ranks of somersaulters—they looked like martial pinwheels—until the victors finally took the castle by double-somersaulting over its wall. At this, the French audience cheered, it being the finale of the program.

The Chinese People's Republic's Peking troupe was bound to arouse the greatest curiosity this year of all the national productions sent to the International Theatre Festival, since Paris had never seen any Chinese Communist actors before, and their country is a crux in the political news. The troupe's shabby, colorful gaiety and exotic high-style circus entertainment seem to have made the most fetching propaganda of any company so far.

A hundred and thirty paintings by Picasso, constituting the most significant group of his works exhibited here since the famous Galerie Georges Petit show of 1932, has opened at the Pavillon de Marsan, in the Musée du Louvre, and will be on view through October. A parallel exhibit of his etchings, drawings, book illustrations, and lithographs will open this week at the Bibliothèque Nationale for the same period, so that people, and especially tourist visitors, can see all that is to be seen of him in Paris at one fell swoop. This is the highest honor possible for the French state to render to a living artist—the privileged invitation to have his works on view concurrently in the nation's two greatest treasure buildings. Everyone important in the Atlantic world as a Picasso connoisseur, collector, or museum expert, or as an authoritative Picasso writer has been involved in putting his show together. The names listed for thanks in the catalogue compose an international modern-art *Who's Who*—people in Paris and elsewhere in France, in England, Sweden, Switzerland, Italy, and the United States. This, with the omission of

Moscow and Prague, is just about the itinerary the collecting instinct for his art originally took, after a few men and a woman in Paris, mostly foreigners and certainly not rich, appreciatively sighted his genius here about fifty years ago and bought some of it at what they could afford to pay, which was not much but which he was glad to get then. Today he is the highest-priced living painter, for both his current works and his old works fetch the highest modern-art prices on earth. An early owner of a fine Picasso portrait lately sold it for the equivalent of twenty-two thousand dollars to a dealer, who shortly resold it for double that.

Among other intelligent, unusual, and sensible bits of personal information about the painter in the Pavillon de Marsan catalogue is the statement that Picasso's *point* today is a hundred thousand francs, a *point* being a unit of measure in the artist-dealer milieu, according to which, for instance, a thirty-*point* Picasso canvas, or one measuring nearly thirty six inches on its long axis, would have a minimal asking price of three million francs—something in the neighborhood of $8,570. However, the selling price, if Picasso chose to sell, would naturally be much higher. According to his art dealer and old friend, Henry Kahnweiler, who was one of the earliest Cubist authorities, Picasso never sells this year's pictures, because he is developing what is in them, and usually keeps last year's pictures, too, as points of reference until he is sure he is through with what they represented in his work. He also never sells personal pictures, of his children or of his mistresses, and is shocked if they sell a portrait he has painted of them and given them, for, being punctiliously Spanish, he does not regard such art as publicly vendible. The most recent sale, M. Kahnweiler says, was to the Swiss Kunstmuseum, of Basel, which bought "Le Bord de la Seine, d'Après Courbet," painted in 1953, for a little over ten million francs, or around thirty thousand dollars. This is one of the amazing series of experiments Picasso has been making in reworking the composition of a previous great painter in his own style of invention.

Though the Pavillon de Marsan show furnishes a magnificent retrospective of Picasso works from 1898 through 1953, many the property of the artist and never exhibited before, they represent a long promenade over fairly familiar rich ground. The *clou* of the show is the fourteen Picassos called "Les Femmes d'Alger," which are variations, in this new manner, on a Delacroix picture of that name, painted in 1834 and now in the Louvre. Picasso started his series on December 13, 1954, and finished it on February 14, 1955. The theme began with two Oriental women, one smoking a narghile, and a servant holding aside a curtain, in a haremlike room with a latticed window. It is impossible to describe the alterations he has made in it, for they are like the slight shifts a kaleidoscope makes when it is turned slowly, its colored fragments evolving new patterns before one's eyes. Fourteen of these new patterns Picasso has halted and fixed in his fourteen pictures, occasionally adding one more woman and then subtracting her again as a radical change, or making one figure recumbent, as if drooped in slumber, and then elevating her limbs into the air, with a few of the pictures in low gray colors and the others resonant in scarlets and whites. What these fourteen variants amount to are fourteen different Picassos, in a new form of multiplication of his own art and style, with similarities separated by his genius. Already appreciations—and, with them, prices—are beginning to form in the air around "Les Femmes d'Alger" as an extraordinary multiple unit that illustrates his styles—a unit that, it is supposed, must go to a museum intact. Probably only a museum, as a multiple unit itself, could afford it.

June 22

This is the end of the 1955 Grande Saison, passed in a mixture of soft sunshine and chilly storms amid a climate of French prosperity, discernible in Parisians' being able to afford their own high prices. There have been elegant dinners, soirées, and entertainments, and the elaborate weddings proper to this annual high social season. On the Paris Bourse, the overbid stock market took a natural slump, but the French class that has money seems to have a lot of it, and to be spending it. France's Armée de l'Air and airplane industry enjoyed the luxury of a great satisfaction this week, just ten years after they started with nothing, at the end of the war—their impressive participation, along with nine other nations, including the United States and England, in their own XXIe Salon International de l'Aéronautique at Le Bourget Airfield. Last Saturday and Sunday, a weekend crowd of half a million saw France's leading trio among its many new planes—the jet bolide called the Trident; the jet transport called the Caravelle, capable of carrying nearly a hundred passengers at a speed of close to five hundred miles an hour; and a helicopter nicely named the Alouette, or lark, which has risen higher in the air than any other of its genus. Luxury hotels are so crowded with tourists, mostly Germans and Americans, come to see the June goings on, that families are sleeping three to a room. Among the myriad attractions was the reopening of Florence's, the famous Montmartre champagne night club; the Grand Palais international horse show; racing at five tracks; Bois de Boulogne polo; Basque pelota; and a rose growers' competition at Bagatelle, won by an American rose fancier. The suburban Casino d'Enghien offered, along with its gambling tables, Verdi's "Le Bal Masqué," with La Scala singers, including Tagliavini. At a televised concert of sacred music, at the Ste. Chapelle, with floodlighting on the thirteenth-century stained-glass windows, an American Negro troupe assembled and led by that folk-song artist from the Left Bank, Gordon Heath, sang spirituals. The Brazilian composer Heitor Villa-Lobos, increasingly popular here, directed the Orchestre National in a program of his own works, among them a first performance of his new piano concerto. In a welter of art shows, there have been such items as an exhibition of Hiroshige prints and drawings, an exhibition of seventy Rembrandt drawings at the Louvre, and the Galerie Louis Carré's highly appreciated show of twenty-seven paintings, dated from 1913 through last year, by the seventy-nine-year-old Jacques Villon, still semi-Cubist, still as gifted in his belated fame as he was before the First World War, and even before the Second, when no one paid any attention to him. The Grande Saison ends this week in the Grande Semaine and the Grand Prix de Paris at Longchamp, run as always on the last Sunday of June, the most fashionable sporting Sabbatical race of the year.

It should be added that, on top of everything else going on in Paris, the city is unfortunately being improved, so that automobiles can have an easier time. Part of the Boulevard Raspail's fine central allée of trees is being deforested to make a shadeless, ugly single-track traffic artery. The magnificent chestnuts by the Place de l'Alma are gone, victims of a riverside traffic tunnel under the square, which looks as though a bomb had hit it. The Avenue de l'Opéra's sidewalks are being sliced away a metre and a half on each side—or about ten feet in all—so that cars may have more room and pedestrians may walk all over each other. Now the energetic Prefect of Police, André Dubois, who at least made Paris automobiles silent by forbidding horn tooting, which everyone said would be impossible, has a plan for garaging them underground, which sounds all too believable. The Municipal Council is about to ask for bids for his *urbanisme-souterrain* project of building a three-story parking garage for eighteen hundred and thirty cars under the Tuileries gardens; eight others at various points, including the Champs-Elysées; and a five-story garage for six hundred cars beneath the Square Louvois, practically in the cellar of the Bibliothèque Nationale. Altogether, these would shelter an estimated eleven thousand cars. The Municipal Council also intends to build thirty-eight hundred small Paris apartments for families of three, or shelter for about eleven thousand four hundred citizens. It is nice to see that in the modern housing struggle here between people and automobiles, the human beings are a little more than holding their own.

The Salon International de la Police has put on a policemen's show called "Le Faux dans l'Art et dans l'Histoire," devoted to the topic of counterfeiting in art and history, which deserves special mention because it is so odd and interesting. Nearly everything on display is false by wicked intent to defraud, which is why it fell into the police net—often from the hands of collectors who got fooled and complained. To begin with, it shows half a dozen patently counterfeit smirking Mona Lisas. To add to the fascination of this show, ranged upstairs in the Grand Palais, original, true works are occasionally included with the black-hearted counterfeit jobs to educate the visitors—most of us being unable to tell the difference even when it is demonstrated right under our noses. The disparity of the things falsifiers have falsified is incredible, for they are connected only by their unifying logic as deceptions concocted to get money. For instance, there are false laces—old Chantilly, Valenciennes, and so forth—that must have been as much trouble to fabricate as the originals. There is a whole library of false handwritten manuscripts and letters collected by a simple-headed French Academician—a letter from Christopher Columbus to Rabelais, one from Napoleon's little son, and even one from Sappho. There are cases of false money, which the police at first hesitated to put on view for fear of giving encouragement; false stamps; false German food-ration cards from the war; bogus antique Greek drachmas; and examples of the handiwork of the master forger named Becker, of the last century, whose spurious coins of Plautus's time, in the second pre-Christian century, are beautiful copies of the hand-struck originals. False pewter objects abound—an industry cooked up under the Second Empire, when Renaissance goblets and plates became a fad among the *nouveaux riches.* The faked pictures are, naturally, among the most interesting exhibits, especially when accompanied by museum laboratories' radiographic photos showing the deceits, increasingly difficult to practice today. The two greatest art fakes shown are famous false Vermeers made by that Dutch genius van Meegeren, who sold his works for fortunes during the war to a Dutch museum, as well as to the far less expert Hermann Göring. Yet "Christ with Martha and Mary" looks faked even to a layman, and its explanatory radiograph shows up a battle scene painted earlier behind the two sisters. But a second, smaller van Meegeren—of Vermeer's model in that familiar yellow jacket—is a slick, lovely job. Also displayed are a fake Rembrandt and a fake del Sarto, fake Italian primitives, dozens of fake late Renoirs, when his brush had grown loose and greasy, and fake Courbets, Corots, Degas, and Toulouse-Lautrecs. Historically, the greatest item offered is the Esterhazy *bordereau* of military secrets that led to Captain Dreyfus's arrest and started *l'affaire.* There is also a photostatic copy of Colonel Henry's infamous forged letter inculpating Dreyfus, the original of the Colonel's shameful confession, the letter from poor, honest Picquart that helped unravel the mystery, and many other engrossing period documents assembled from this counterfeiting of guilt that cut France in two fifty years ago. What the international police have put together in "Le Faux" is a highly interesting, cynical exhibition of human credulity and beguiling deception.

July 6

The most acclaimed theatre item in the Festival International d'Art Dramatique, at the Théâtre Sarah-Bernhardt, was unquestionably Bertolt Brecht's Berliner Ensemble, from Communist East Berlin, in his piece called "Der Kaukasische Kreidekreis," or "The Caucasian Chalk Circle." Its peculiar title refers to a chalked circle supposedly scrawled in a market place somewhere in the Caucasus, around which two women struggle before a judge for legal possession of a child in the center, each trying to pull him out of the circle to her own side. The honest Georgian peasant woman who nurtured the boy after his wicked, rich, noble Persian mother deserted him is judged the truer mother, and wins him because she chooses to cede him rather than pull at his arms as if pulling at the wings of a fly. The final morality song that accompanies the curious play gives the dialectic key to this Sovietized version of King Solomon's judgment: "Each thing belongs to the one who does it the most good." The rest of the play is far more complicated, being set during a medieval war between honest, bare-faced, ragged Georgian peasants and evil, wealthy, masked and costumed Persian nobles, and involving a dozen-odd characters, the passage of several years, many journeys, and some entirely independent folk stories full of gusty, lusty repartee—all this overrich content for the ear made coherent through its presentation as a kind of vision that gives its news direct to the spectators' astonished eyes. Brecht's play was indeed something to see—a stylized, *maniériste,* modernist stage production, *echt deutsch,* yet certainly derived from the early Moscow experimental theatre of Taïrov. The big, bare, white-hung stage was decorated on the left side only, by a severe backdrop—a succession of pendent strips of white silk painted with Chinese-looking black landscapes, constantly renewed to indicate change of scene. The peasant woman, with the boy in her arms, sometimes plodded on a treadmill to indicate her journeys, and sometimes little gray huts rolled toward her on its track to offer shelter. As for Brecht's noted theory of *Verfremdungseffekt* in acting, according to which his actors are supposed to be physically detached from their roles, so as to leave the audience's precious critical faculties free, it seemed merely to boil down to excellent stage directing, which permitted no emotional ranting and featured cold realism almost to the point of symbolism, and thereby stirred Parisian critical faculties to appreciation, if not comprehension. Accompanying the stage performance were three singers, seated on the proscenium, whose songs, to music by Paul Dessau, insufficiently recalled the famous topical manner of Brecht and Kurt Weill in their "Dreigroschenoper," the work that first made both of them internationally known, almost thirty years ago. Brecht and his wife, Helene Weigel, have organized what seems the most stimulating experimental theatre of Europe today. He appears to be the playwright most intimately troubled by the events of our epoch, and he has projected his disillusioned imagination onto the German stage as his form of relief.

It should be kept in mind that the works presented here in the Paris drama festival by almost every free nation in Western Europe, plus Fascist Spain and Red China and Iron Curtain Poland and Communist Berlin, all functioned as artistic propaganda vehicles, like the more elaborate contributions in the American "Salute to France."

July 20

For the first time since the Liberation, July 14th seemed a real *fête nationale*—a day for public pride, merrymaking, and the parading of soldiers. During the past ten years, the old July victory of the Bastille has been superseded by the June defeat of 1940 as an annual recollection, and simulated martial pomp marching down the Champs-Elysées has been a spectacle of humiliation for French eyes to look upon while French memories were helplessly remembering. Maybe a decade and a half is a span of time in which agonizing history can fade. Anyhow, throughout this present year the French seem to have been turning some invisible psychological corner and to have rediscovered themselves. At the start of the Bastille Day parade last week, cheery applause rose from the perspiring, shirtsleeved sidewalk crowds for the President of the Republic, René Coty, standing upright in his open car, and wearing evening clothes at nine-thirty in the morning, with his scarlet *grand cordon* of the Légion d'Honneur flashing in the heat wave's early hot sun, and there was also vigorous hand-clapping for the St.-Cyr cadets, in their ancient tricolor uniform—white cassowary plumes, blue pinchback frock coats, ballooning red pantaloons.

Painful present-day history also had a new treatment. France's relations with her North African possessions still being at the point of bloody insurrection, some carefully chosen colorful native troops and black majesties from the further reaches of the French Union had been summoned to appear on Bastille Day as honored participants and guests, the parade's spectacular novelty being the Garde Rouge from Dakar, riding small Arab stallions. These handsome brown men made a barbaric picture, like a group by Delacroix, transported to canter down the metropolitan avenue in their tall red *chéchias,* flowing red capes, and blue Oriental trousers, with drawn scimitar-shaped swords. There were also marching youths from military preparatory schools in the Sudan and on the Ivory Coast—the educated young generation, solemn and unpicturesque in

European white gaiters and khaki uniforms. Sitting among the diplomats in the President's reviewing stand were amazingly garbed dark visitors, the cynosure of all eyes—especially His Majesty Alohinto, King of Dahomey, wearing an incongruous gold-embroidered French Army kepi above his draped bright silk robe (but without his fancy fringed umbrella, which had been held over him when he was being presented to Mme. Coty in the Elysée Palace). The young Sultan Youssouf, of Massénya, in Chad, wrapped in white wool, wore a dignified little white cap, but the African hinterland chiefs were more savage in fancy headgear, with floral embroidery, and festooned, multicolored gowns.

The other novelties in the Champs-Elysées celebration were strictly modern—the new, top-grade French military equipment, including powerful, fast tanks, to show France's recent recovery, and, for the first time, public formation flights of the latest French jets, led by forty-eight Ouragans, screeching high in the air as they disappeared over the Louvre and followed by twelve big Mystères. The traditional night dancing in the streets seemed newfangled, too, with few *bistro* accordions breathing sentimental waltzes. Generally, they had been replaced by phonographs playing mambos. One of the grander *bals du quartier* centered around the firemen's *caserne,* a stone's throw from the Ritz, in the Place du Marché-St.-Honoré. The unusually artistic firefighters who occupy the building had draped their emptied apparatus room with painted curtains representing the old walls of Paris, had hung the ceiling with highly inflammable tissue-paper decorations, and had set up their bar in the doorway from which the hook and ladder usually emerges. The dancers were mostly the neighborhood marketmen and their families, professionally indifferent to occasional whiffs of fish rising from their adjacent stalls. It was a very animated ball.

The day the Dakar Garde Rouge was galloping as a guest in Paris was Casablanca's worst day of anti-French riots. The festering Moroccan and Algerian problems are still to be settled. Though plenty of anti-home-rule deputies had previously said the fate of France, or at least of Tunisia, was at stake, only about a fifth of the six hundred and twenty-seven deputies bothered to turn up for the opening Assembly debate on ratification of the Franco-Tunisian home-rule pact. The surprisingly large majority that finally ratified it was created by the Communists' last-minute switch, when they gave their ninety-eight votes to what they cheerfully called "the first step in liberation of an oppressed people" (though the liberation of the Communist-satellite oppressed peoples is an item the Russians have wanted to hear nothing about at Geneva). In the debate, the anti-home-rule Gaullist splinter party, the Action Républicaine et Sociale, accused Premier Faure of holding somebody else's baby, meaning that Mendès-France fathered the pact last year when he declared at Carthage that France would really start living up to her promises of reform—made years ago, and always side-stepped till now. Clever, elderly Paul Reynaud begged the deputies to be "men of 1955," meaning that colonialism is over, as, indeed, Premier Faure also stated, adding that if the Tunisians hoped for eventual independence, who could "deny this ideal," ironically learned from French libertarianism? Experts see the ratification of the Franco-Tunisian pact as a turning point in the history of the French Union—France's first surrender of power in her African territories. Since 1951, Parliament had refused to cede an inch to Tunisian nationalist aspirations. The recent change of mind may have been pushed by French public opinion, weary of bolstering French authority by force and risking another Indo-China.

Eleven large and lovely gouaches by Juan Gris, painted in 1917, then put away, and until now unknown to the public, have finally achieved the book form for which Gris intended them, as Cubist lithographs illustrating Cubist poems written in 1917 by Pierre Reverdy, who today is regarded as one of the half-dozen great living French poets. The volume has just been published, under the title "Au Soleil du Plafond," as a bibliophilic item by Tériade, editor of *Verve* and probably the only Paris editor who would lovingly risk such a refined enterprise. It took three years to prepare the volume. There are two hundred and five copies, priced at forty-five thousand francs—or about a hundred and thirty dollars—each, and worth it to collectors. The story behind the book is an odd one. According to Reverdy, in 1917 Gris and he conceived the idea of this doubly Cubist book, to contain twenty poems, which Reverdy duly wrote, and as many gouaches, of which Gris painted only eleven. The volume was accepted for publication by the art merchant Léonce Rosenberg, but the project fell through for various reasons—maybe partly because Gris was not appreciated until long after he was dead. Picasso and Braque were the Cubists who were then bought by intelligentsia collectors—when any Cubists were bought at all. Gris was lost in their shadow; his pictures started to sell only just before this last war, and then sometimes for as little as a few hundred dollars. He was, Reverdy says, a melancholy Spaniard who loved to drink coffee and dance at the artists' balls at the Bal Bullier. He was also an exile, unable to go home because he had refused to perform his military service, so when, in the mid-twenties, he felt his health was failing, he had to send to Spain for his son, Georges González Gris, who had been brought up there, to come and say farewell. Gris died in 1927. The son kept the eleven gouaches as his father had left them—in an old cardboard box. Today, the son is a chemist in a French factory that makes paint for automobiles, and is not interested in art. Three years ago, he suddenly wrote to Reverdy, asking if he knew anything about those gouaches. The present book is the result.

The eleven original gouaches are not for sale. In the book, their pure Cubism of the *belle époque* seems intact, as if exhumed in a perfect state of preservation. They bring to life once more the severe iconography of the period—the violin, the sheet music, the lamp, the bottle, the pipe, the fruit dish, and the coffeepot, related by the imagination of Gris in browns and grays, sometimes with a core of purple or blue, and occasionally with dapples of white. Reverdy saw with the same eyes, achieving in words the same strict, methodical sobriety and the same emotional essence, wrought deliberately from the commonplace. His poems appear in the book in his handwriting, lithographed, like the Gris *estampes,* by Mourlot Frères, master lithographers of Paris. In its union of poetry and painting—especially coming more than a third of a century after its native epoch—it makes a rare and interesting volume.

August 3

It is a blessing right now that France's North African holdings are not as vast as French imperialist visions used to be, because the bloody troubles in Algeria, Tunisia, and Morocco are more than the government in Paris can control, or the more moderate North African Nationalist leaders can calm, or the roused Moslem mobs can resist expanding into further bloodshed. The state of emergency in Algeria, which is in its tenth month of armed rebellion, has just been prolonged, on general principles, to next spring by Parliament. This week, Tunisia is to enter into the uncertainties of its limited home rule after more than three years of sporadic bloodletting. For the past fortnight, Morocco's events have daily filled the Paris newspapers with awful reports and awful photographs from former favorite tourist and residential cities like Marrakech, Casablanca, and Rabat, where there has been death and destruction in cafés, cinemas, and streets, with white counter-terrorists killing both whites and natives, and natives killing whites and one another—especially those favoring the whites—in what seem outbreaks of savage incoherence. The French government's only sustainers of phlegm and order appear to be the Foreign Legion, with its helicopters and tommy guns, and M. Gilbert Grandval, the new liberal French Resident General for Morocco, imperturbably shouldering his way through rioting mobs, both brown and white, on his tour to discover what is going on down there and why.

The most prodigious summer exhibition is in the Château de Versailles. Entitled "Marie-Antoinette—Archiduchesse, Dauphine, et Reine," it is the richest evocation of her life ever assembled, and will be on view until November 2nd, the bicentenary date of her birth. This unsurpassable exhibition was largely brought about by the initiative of two amateurs (the Baronne Elie de Rothschild and the Duc de Mouchy) and one scholarly professional (Gérald van der Kemp, the curator of Versailles). It contains nearly a thousand contributions lent by the few royal families still functioning, by princes, nobles, and courtiers' descendants, by palaces, museums, and private collectors everywhere. All these items bear the burden of history, the stamp of Marie Antoinette's person or her family, her fatal follies and final tragedy. The numerous portraits are of extreme interest, starting with those depicting her on her arrival in France from Austria, when her pleasure-loving blue eyes and maidenly fresh coloring were gratefully seized upon by court artists as compensations for her lack of beauty. Then, after the decade that it took for her to lose the French nation's affection, even the temperate portraitist Mme. Vigée-Lebrun began realistically portraying her heavy, obstinate Austrian jaw. And finally (as the exhibition also shows) came the Revolutionist David's triumphant and terrible little sketch of her, riding backward in the tumbril a fortnight before her thirty-eighth birthday.

Of the exhibition's hundreds of intimate objects connected with

Marie Antoinette, and miraculously saved from the Revolution's fury, a special few frivolous or tragic items seem perfectly to sum up her life and time. Among them is the early warning letter from her mother, Maria Theresa, scolding her for her dangerously silly conduct—"more like that of the Du Barry" than the behavior proper to a young archduchess. Another illuminating sight is the Sèvres *bol-sein,* or breast cup, modelled from her bosom when she was Queen. From it, guests supped milk while she played at being a dairymaid at the Petit-Trianon. But the pièces de résistance are her court jeweller's copy, in crystal, of the famous *collier de la reine*—the mysterious, fabulous diamond necklace that, involved in a public scandal, helped overthrow the extravagant monarchy. Twenty-two of the original huge white diamonds, which were smuggled to London, have been lent by their present owner, the Duchess of Sutherland, and are on view in a simple setting. Also shown are Marie Antoinette's sapphire-and-diamond jewelry set, which was purchased after the Revolution by Napoleon for Joséphine, and has been lent now by the Comte de Paris, heir to the theoretical French throne.

Among the enriching background contributions are two unroyal portraits—one of the Queen's rather untalented dressmaker, Rose Bertin, and the other of the writer Beumarchais—and Her Majesty's copy of Beaumarchais' "Le Mariage de Figaro," which she had attendants read aloud to her at the very time when it was fomenting uprisings in Paris. On loan from the present Queen of England is the French Queen's wonderful trick clock, La Négresse—an automaton that gave the time by rolling its eyes, and also could play tunes in its black head—and there is a letter from a watchmaker named Gide (of the family to which André Gide later belonged) about repairing the Queen's so-called perpetual-motion watch, which she had dropped. The last folly displayed is the silver-fitted travelling case that was specially made for her to carry on that vain, clumsy Varennes flight from the mob. From her days as prisoner in the Temple, there is shown a small waistcoat that she embroidered for the little Dauphin—pale flowers on mourning-mauve silk—during her dreadfully idle hours there, and also the wooden spoon furnished for her meals. The final exhibit is part of a black silk stocking that she wore to her execution. It was exhumed, with her remains, in 1815.

Far more than her lethargic Bourbon husband, Louis XVI, the Austrian Marie Antoinette still represents to the French the last of

France's *ancien régime,* and her fall the beginning of modern times. The impressive Versailles exhibition being the first of its kind, Paris papers have run serious articles asking whether it will make today's Fourth Republic children love or hate her. Nancy Mitford, who is a resident of Paris, reported on the exhibition for the London *Sunday Times* and stirred up a hornets' nest by writing, "To me, Marie Antoinette is one of the most irritating characters in history. She was frivolous without being funny, extravagant without being elegant; her stupidity was monumental." Though she duly added, "When all hope was gone, she became an exemplary figure," not only were there protesting letters to the *Sunday Times* but she herself received hundreds from the French (from indignant vicomtes, from pious, elderly provincial royalists, and also from obviously unreconstructed Vichyites), oddly deploring "the fate of our dear Queen"—small, if significant, leftover contributions to France's confused political tableau of today. According to history, Marie Antoinette's last remark was to her executioner, on the scaffold, when she accidentally trod on his foot and said, *"Monsieur, je vous demande excuse. Je ne l'ai pas fait exprès."* ("Excuse me, sir. It was not intentional.") As *Le Figaro Littéraire's* critique on the Versailles exhibition concluded, it was an apology that she might well have made to the entire French nation.

August 17

The Bibliothèque Nationale's exhibitions built around great French writers from the past are noted for their exhausting effect on the visitor, the material being so complete, stimulating, and irresistible that it takes hours to see it. The present exhibition, honoring the bicentenary of the death of the Duc de Saint-Simon, France's amplest secret memorialist and most illuminating historical gossip, is even more special. On view for the first time in a century are the eleven elegant calf-bound portfolios of his original "Mémoires," written in his swift, legible script and adding up to forty-three volumes of print in their complete modern edition, which true Saint-Simonites, by skipping the dull patches, still read like a serial whose satisfying end they already know. Among the rich memorabilia, the Bibliothèque features his famous portrait, showing his wide, observing eyes and doglike short nose, tilted to catch the scent of news, along with portraits of those whose news he secretly wrote about, such as old Louis XIV and his frigid Mme. de Maintenon. A small, frail nineteen-year-old ducal officer off at the wars, he hurried in from a Rhineland post one evening, sat down in his tent, and suddenly began writing. From then on, he kept notes on everything and everybody, and at forty-eight he started refining them into his memoirs ("I was born in the night of the fifteenth and sixteenth of January, 1675, of Claude, Duc de Saint-Simon, peer of France, and of his second wife, Charlotte de l'Aubépine, only child of this bed.") He died at eighty in his rundown house in Paris, in debt for the candles used for his night-and-day writing, disgraced, and dismissed from the court at Versailles, where he had lived for thirty years, summing up its intrigues, its history, and its dominant personages; covering the elderly Louis XIV, the Regency, and the child King Louis XV; and mentioning seventy-three hundred and fifty people by name, often devastatingly. He had a fetishist's worship for court etiquette—such as the precise angle for a noble's bow—revered chastity, unpopular at the time, and genealogy; organized and wrote about his dangerous campaign against the King's legitimatized bastards; was against the King's persecution of Protestants and against his Ministers' bankrupting taxation of peasants (and dukes); and was received only three times in royal private audience. The King told him he talked too much, having no idea that what he was writing would be far worse.

Chateaubriand, Stendhal, and Sainte-Beuve were the first modern writers to appreciate the "Mémoires," which eventually influenced Proust, who applied the Duc's passion for etiquette and genealogy to the Guermantes, and took something of his overloaded syntax and style. Today, the "Mémoires" are still appealing, because, along with events, grave or gossipy, Saint-Simon supplied realistic, scintillating word portraits, the first in French literature. Of Mme. de Castries he wrote, in part, "She was a quarter of a woman, a kind of unfinished biscuit, very small . . . with no backside, bosom or chin; very ugly. . . . She knew everything—history, philosophy, mathematics, and dead languages . . . was cruelly malicious and polite." And of the Prince de Conti, he wrote, "This highly agreeable man loved nothing. He had and wished to have friends as one has and wishes to have furniture." Illness and deformities fascinated him, and he reported on all aspects of them—on Pontchartrain's glass eye, on the Duc de Bourgogne's crooked shoulder, on the Duc de Vendôme's syphilis, on the Chancellor's dropsy, on one royal bastard's indigestion and another's apoplexy, on the royal mistresses' miscarriages, and on the King's fatal gangrene.

Some new books on Saint-Simon have just appeared in connection with his bicentenary, but the best seems still to be "Saint-Simon par Lui-Même," by François-Régis Bastide, which appeared year before last. It is one in a satisfying cheap series of generously illustrated pocket books on great French writers. They are written half *"par lui-même"*—that is, by the great writer himself, through quotations from his works—and half by the modern critical commentator, who in this case is lively, amusing, and expert. The little Duc having been the first gossip columnist, it seems strange that Bastide's volume has not been put into English for Anglo-Saxon gossip column livers of today.

August 30

Last week, after years of French inattention to swelling Moroccan political demands, French bullets furnished the first rapid reply Moroccans have ever received from a French government. In repressive measures carried out by the military governor of Casablanca, a thousand natives, more or less, were shot around Oued Zem, in classic punishment of roughly ten to one for the eighty-eight white French colonials barbarously massacred in and around that settlement during the horrible uprisings that started on August 20th. *"Au Maroc, Messieurs, le temps c'est du sang"* ("In Morocco, sirs, time means blood") was the warning earlier given the French by Gilbert Grandval, Premier Faure's liberal Resident-General there. To his forceful mind, haste was finally vital for the French if they were to work out a rational accord with Moroccan Nationalists that could both protect French interests and forestall the sanguinary, fanatic violence otherwise due on August 20th. August 20th was the anniversary of the day two years ago when the then Foreign Minister, Georges Bidault, let the pro-native Sultan Sidi Mohammed ben Youssef be deposed and bundled off to exile in Madagascar with a small selection of wives. There ben Youssef became for the Moroccans what Habib Bourguiba, exiled in France, had become for the Tunisians. The ex-Sultan, a fairly worldly Moslem, was transformed into an anti-French martyr and national symbol—something the Moroccan independence movement had lacked until the French ironically supplied one, to be used as a wild rallying cry for this year's significant August 20th anniversary. After the installation of his successor—an obscure, elderly pro-French puppet, Sultan Sidi Mohammed ben Moulay Arafa, regarded as illegitimate by most Moroccans and especially by the Istiqlal, or Independence, Party and its followers—Moroccan affairs worsened to the point where Faure, who shared Mendès-France's North African reform ideas (for which Mendès-France was thrown from power by Parliament, whose majority deputies then paradoxically chose Faure to succeed him), sent Grandval to Morocco to make a first-hand report, complete with advice. Grandval's proposals practically duplicated Mendès-France's. And the powerful anti-reform colonial diehard interests and their lobby of conservative deputies, who had defeated Mendès-France and his projects, duplicated their success by defeating Grandval and his plan. Even after the fatal August 20th that he had prophesied, his reform policies were still unacceptable last week to the section of Faure's Cabinet that misnames itself the Moderates. (No French political group today ever calls itself conservative, let alone reactionary.) To save his government from falling in defeat in its turn, though its usefulness and reputation are already much weakened, Faure has just offered to trade a couple of heads—the head of Grandval, who had already offered to resign as Resident-General, to placate *les Modérés,* in exchange for the turbaned head of Sultan ben Moulay Arafa, supposedly ready to oblige by abdicating, to placate Moroccan Nationalists, who demand the return of Sultan ben Youssef. But old ben Moulay Arafa, chosen because he was presumably so malleable, has twice declared that he will not give up his "divine mission" as head of the Moslem faith. This has been a shock to the conference of French and Nationalist leaders now going on at Aix-les-Bains, since the throne, as the French have discovered too late, is the spiritual and temporal keystone of Moroccan unity. There may be another shock, and possibly another bloody upheaval in Morocco, when the liquidation of Grandval is formally proclaimed to the natives. He is the Frenchman they have learned to trust.

In the interim, whatever happens this week in the way of a Franco-Moroccan accord will be too little and too belated. As *Le Monde* severely stated, the Moroccan problem, "at the rate it has been allowed to rot away," can result now only in some sort of compromise, which will be "neither glorious nor sufficient." Parisian journalists and cameramen sent down to cover the sudden Moroccan campaign (a reporter and a photographer for *France-Soir* and an N.B.C. television-movie photographer were murdered in the uprisings) furnished descriptions and photographs of the anomalies of Moroccan life that were news to many French. In text and pictures they showed the modernity of the French cities and towns, with their up-to-date whitewashed concrete, and, outside them, the anachronistic native Moslem rural tents, made of tree branches and shaped like wigwams. Barrett McGurn, covering the uprisings for the Paris edition of the New York *Herald Tribune,* gave Americans here the most effective picture of those brush habitations from Morocco's nomad past. It is, he said, "as if the Indians had stayed on in New York, still maintaining their ancient tribal ways, still brooding about scalping parties, never merging into the modern world of subways and automobiles."

The famous creators of modern painting in the Ecole de Paris, having grown old—as, indeed, modern painting has itself grown old—are disappearing. First Derain, then Matisse last fall, and now Fernand Léger are gone, leaving behind only Braque, Rouault, and Picasso. Léger, always an active, resolute man, died quickly of a heart attack, aged seventy-four, in his country studio outside Paris, in the valley of Chevreuse. As a young man, he early came under the influence of Cubism, which he turned into a version irresistibly called Tubism. In the First World War, he served with the artillery, designing breechblocks for guns, and he was thereafter obsessed with the cylindrical shape of gun barrels. Fascinated by the constructive aesthetics of machinery, he not only applied the patterns of mechanics to his compositions, which frequently featured bicycles, but mechanized the human figure to match. He was the one major French artist who was influenced by, and found himself perfectly at home in, the United States, and during the last war he lived in New York, where he found "the glorification of the machine." (As early as 1924 he had made a curious, interesting movie called "Le Ballet Mécanique," which starred the movements of pistons and other pieces of machinery.) Back in France after his American period, he contributed a huge gay mosaic, interpreting the Litany of the Virgin, to the famous modernist village church of Assy, in Haute-Savoie, and he recently designed a brilliant series of stained-glass windows fo the church at Audincourt. Léger's beautification of these churches which were also decorated by other leading modern masters, was part of the aesthetic scheme originated by the late Père Couturier, the intellectual Dominican who wished to use the greatest living artists for the Church's glory—regardless of what some worshippers called their unholy style—just as the great Italian churches used artist geniuses in the Renaissance.

An extraordinarily sober evaluation of Léger by the Paris art critic André Chastel has just appeared here as an obituary—one that might well make the artist turn in his grave. Chastel said, in part, "From 1910 on, his views of cities with smoke like zinc, his country scenes incised as if by a wood chopper, his still-lifes made as if of metal, clearly showed what always remained his inspiration: the maximum hardening of a world of objects, which he made firmer and more articulated than they are in reality. Sacrifice of color and nuance was total, and line was defined with severity and a well-meaning aggressiveness, reflecting his violent, cold, Norman temperament. This revolution he consecrated himself to seemed rather simple—the exaltation of the machine age, which, after 1920, dominated the Western world. . . . To some huge canvases of perfected articulation, painted between 1925 and 1930, Léger brought a tonic richness that summed up the century. . . . Any definition of his art, whose energy and tension were based entirely upon the decision to be modern, makes clear his art's limitations." Léger's funeral, held in his village studio, where a half dozen of his Tubist pictures, on easels, were placed like mourners behind the flowers around his bier, was held under the auspices of the Communist Party, of which he was a member. The funeral oration was given by Etienne Fajon, secretary of the Party, which, Fajon said, Léger had "loved with all his heart and served with all his might."

There seemed no special rush about reading the now completed first two volumes of Paul Léautaud's "Journal Littéraire"—it may eventually cover fifty years of his diary notes—since the first volume starts with 1893 and the second, published a few months ago, ends with 1909. The author, now eighty-three years old, was born the illegitimate son of a prompter at the Comédie-Française; as a little boy, he sat in the prompter's box, soaking up the French classics, and at thirty-five he became what he remained for many years—the ill-paid, rather obscure drama critic of the well-known Paris magazine *Mercure de France.* He burst out as a public figure five years ago, when someone had the bright, dangerous idea of putting him on the radio for weekly reminiscences about the homeless cats he has loved and collected in his life. His interpolated reminiscences about his former mistress of twenty years, whom he called *"le fléau"* ("the scourge"), were so pertinent that he was cut off the air, and censored records were made in advance for the rest of his causeries. Piquant selections from his diary, describing amorous quarrels, and especially reconciliations, as late as his sixtieth year, all related with the egotism and candor that he chose as his literary standards, were then suddenly featured in such leading Paris monthlies as *La Table Ronde, La Parisienne,* and *La Nouvelle Revue Française.* His démodé libertinage, his sharp, intelligent pen and tongue, his poverty, cats, steel-rimmed glasses, pixie face, and crowning antique tweed hat belatedly made him a temporarily popular, cross-grained old star in the drawing rooms of Paris intelligentsia, and Cartier-Bresson photographed him, articles featured him, and an issue of *Le Point,* a provincial intellectual magazine, devoted an entire number to him. Not long ago, when an old friend of his was finally being received into the Institut, Léautaud, who had no invitation, turned up for the fashionable ceremony in that hat and a pair of carpet slippers, and carrying a string bag filled with raw meat for his cats. He was refused entrance, and, to the delight of the journalists assembled for the Institut event, gave his uncensored opinion of academic honors generally, some of which could even be printed in the next day's newspapers. For years he has lived, for his cats' sake, in an old house with a garden, now his animal cemetery, in the suburban town of Fontenay-aux-Roses.

Unwittingly, he gave a key to the interest his journal has today for the literary French by early noting in it "the slavery of writing down one's daily ideas and happenings." The result is a serialization of a life and a period that both ensnares and alarms present-day readers. His close friends, if he really had any—his long poverty made him gauche and timid—were Rémy de Gourmont and Marcel Schwob. His acquaintances included Pierre Louÿs, Catulle Mendès, Jarry, Péguy, Mirbeau, Redon, and François Coppée. His chief enemy was Huysmans. In 1903, Léautaud published a little masterpiece of a novel, "Le Petit Ami." It went on sale the same day as Zola's book "Vérité," which by night had sold forty-one thousand copies. "Le Petit Ami" sold less than three hundred in two years, and today is worth a hundred dollars a copy, if you can find it. He loathed Flaubert, calling him "a day laborer of style," and saying that anyone could have written "Madame Bovary" if he worked hard enough. He despised Anatole France as a vulgarian, and adored Stendhal, whose tomb in the Montmartre cemetery he visited to thank him for having written "Lucien Leuwen." Léautaud's own writings scandalize by their candid verbal accuracy only when dealing with his many mistresses, his physical passions, his parents, and anybody's deathbed. A lonely egotist, he wrote in his journal, "My egotism is so natural that I am all that interests me. If I had to write about a table, I would still find the means to speak of me."

November 9

In the past week, Paris has experienced the warmest, sunniest, most beautiful November days since 1899. The only familiar climate has been that of French political storms. Yet even the constantly shifting, bloody insurrectional tempests that swept over Morocco during the last three months, threatening the unity of the North African empire, on which France still sets such store and pride, have now subsided in a peculiar calm, thanks to the return to the Sherifian throne of Sultan Sidi Mohammed ben Youssef, who is right back where he started from when he was exiled two years ago, but with all this summer's burned villages and savage assassinations, bitter mutual recriminations on the part of natives and whites, and incalculable loss of French prestige lying vainly in between, like events that might as well not have happened, since they only circled back to their own beginnings.

The death of Maurice Utrillo from pneumonia, at the age of seventy-one, finally ended his lifelong practice of what could be called automatic painting, a manual reflex of genius that first made its way to canvas through a vinous cerebral haze. He was an alcoholic at the age of ten, and the illegitimate son of a remarkable woman, Suzanne Valadon, a circus acrobat who turned artists' model for Toulouse-Lautrec, Renoir, Degas, and even Puvis de Chavannes. They, recognizing her talent, encouraged her to paint. Maurice was charitably adopted by a chance passerby in her life, Don Miguel Utrillo, who was reputedly a Spanish expert on El Greco. By the time Maurice was eighteen, alienists were advising his mother to lock him up with a paintbox in order to avoid having him shut up in an asylum. Under her home teaching (though her forte was female nudes), in 1902, at the age of nineteen, he began an amazing series of fine paintings in grays, blues, and browns—the visionary pictures of Paris, falsely seen as a provincial town, that were his peculiar invention. These still fetch high prices, as do the paintings of his famous white period, dating from 1908, in which the dark cobblestone streets of Montmartre look purified by snow. In 1921, mother and son held a joint exhibition. By 1924, in spite of almost two decades of intermittent confinement in jails, prison hospitals, and the Asile Ste.-Anne, he had made his name far more famous than hers. There was just enough coarsening of his subtle palette by this time to give him a vital success. His prices rocketed. Over the years, he became the Paris tourists' favorite painter. He painted Paris churches and streets that he had never seen, drawn from postcards, carrying bogus realism into a flight of the imagination. Fake Utrillos appeared on the market, some supposedly painted by the second husband of his devoted mother; his café quarrels were alarming. Sober, he was docile, and said nothing of consequence but painted on order, with an obedient, if lessening, genius. Before his marriage to Mme. Lucie Pauwels, widow of a Belgian banker, his mother and her young second husband became his managers and provided him with fine cars, a country château, and a splendid Paris studio, to keep him out of bars. Since he was valuable and helpless, everyone's effort was to save him. Final salvation and sobriety were arranged for him by his wife, but too late. Only a few months ago, he was still repainting, as if in a daze, his most popular Montmartre scene, the Sacré-Cœur. A unique movie sequence of him at this work was made for Sacha Guitry's forthcoming film, "Si Paris M'Etait Conté." Art critics here rated him, at his apogee of delicate color and nostalgia, as one of France's great city *paysagistes*. Even after he had outlived himself, he was cherished as a mysterious artistic automaton and the last real bacchic bohemian of Montmartre.

Henri Cartier-Bresson is the only photographer who has ever been honored by a one-man exhibition in a national museum here. At the Pavillon de Marsan, in the Louvre, four hundred of his most notable pictures, many of which make up his two new volumes, "Moscou" and "Les Européens," are now on view. They are the first of his works to be published in bound form in France since his famous volume on China in transition, "D'Une Chine à l'Autre," which had a preface by Jean-Paul Sartre. Cartier-Bresson's unfailing historical intuition—which has led him to keep dates, prophetically sensed in advance, with world events and changes, so that he was able to be present to take their pictures—makes his current exhibition a great general contemporary record of our time, sensitively seized by one man's eye and his camera lens. In it, his snapshots, magnified to the size of paintings, become amazing modern portraits of humanity and its background scenes, both private and public, over the past few years, showing what people have done to each other in war, or what has luckily escaped history and remained a perfectly patterned agricultural landscape, such as a Korean rice field or a Greek olive grove; showing continuous proofs of the rebuilding of existence, if only through the presence of Spanish children playing in architectural ruins, or French wine-growers laying down their vintage for the years to come, or Parisian lovers, hand in hand, smiling at the future, or old Russian women praying in church, still planning as far forward as Heaven itself. It is the intimate Moscow pictures, taken last year, before the official affability between East and West set in, that visibly satisfy Parisians' most acute curiosity. The French are not born travellers; Cartier-Bresson has travelled for them, bringing back to the walls of the Louvre what he calls "the decisive moment" of varied national existences, when thousands of people, singly or grouped, unconsciously furnished him, if only for an instant, with those perfect physical compositions which make great pictorial art, and which, with an artist's eye, he seized, and in that instant made permanent.

The battle now going on between ex-Premier Mendès-France and Premier Faure for the leadership of their Radical Socialist Party, which they started splitting in two, each with astonishingly fine speeches, at the Party's Salle Wagram National caucus last week, is really a fight for the leadership of France next year, after the new elections. If effective electoral laws are not passed first, the ungovernable kind of Parliament that sits right now will be sitting then, and, as usual, nobody will lead France except in fits and starts.

December 7

Till now, post-war French governments have collapsed with such ease that their twenty falls in ten years constituted the major connecting parliamentary history of the Fourth Republic. Last Tuesday's twenty-first fall became a harsh crash, through the consequent dissolution of Parliament itself—for the first time since May 16, 1877, a date made memorable by the unpleasant, autocratic Marshal de MacMahon, then President of the Third Republic. After Premier Edgar Faure's government was overemphatically voted from power the other night, there was a long moment of fatigued silence, and no emotion, no hate, no grief—Faure being a chilly sort of leader—such as made the impressive climax to the fall of Pierre Mendès-France and his government last February. That was the occasion on which Faure started his astonishing rise to fame. As the Mendèsist junior Premier, he set up his government to carry on the policies of his fallen chief, and ended it trying to drive him into the wilderness.

In this struggle, Faure has won. Mendès-France—and maybe France herself—has lost. Instead of winning by overthrowing Faure that final Tuesday, the Mendèsists lost through somebody's overshooting—*une erreur de tir,* as it was sportingly called here. Somebody arranged that seven more anti-Faure votes should be cast than were needed for a simple defeat, and thus made the anti-government majority—by six votes—greater than half the total roll of the Assembly. (Ironically, a similar too heavy parliamentary majority overthrew Mendès-France himself ten months back.) Mendès-France is the only anti-Marxist who has ever bothered with youth, the future of France. He calls his new movement the Front Républicain, to distinguish it from the Front Populaire idea, which would include Communists.

After the terrific sensation caused by Faure's dissolution of Parliament, Mendès-France achieved a second-best sensational riposte by having Faure expelled from the Radical Socialist Party. In France, party expulsion leaves a politician as disgraced and homeless as an Englishman who has been pitched out of his club. With all these melodramatic surprises erupting here, the front pages of the Paris newspapers have looked like chapters from serialized thrillers. The episode the Parisians liked best was the Assembly's being booted out, for Parliament is mostly a despised institution today. Among the excitements have been the violent denunciations of Faure's dissolution order as a reactionary plot, an anti-democratic *coup de force,* and a piece of political chicanery that served him rather than the nation for whose aid it should have been intended.

Mendès-France has also been painfully criticized—at least in private—by some of his too exalted followers of last year. He has been reproached by them for the disillusioning, irascible demagogy he has displayed in *L'Express,* the Mendèsists' weekly paper, where, in his brief signed editorials, he has fulminated about "tricked elections," "guilty men," and "a plot against La Patrie." Perhaps the worst of the many ill results of his and Faure's destructive struggle is that the toga of Mendès-France seems to have slipped, or at least to be showing discouraging signs of wear and tear.

December 20

This must be the funniest election campaign—unfortunately reflecting the dire gravity of the whole parliamentary situation—that either of the two modern Republics of France ever saw. It is as if the nation's former genius for old-fashioned farce were finally taking over the nation's politics, which have too long invited it. The fact that the campaigning to elect an entire new Assembly on January 2nd is restricted to a single month makes for a ludicrous situation in itself, and it is complicated by the uproarious confusion of five thousand-odd candidates struggling for five hundred and forty-four deputies' seats on nearly a thousand electoral lists presented by twenty-eight national political parties and scores of minor local groups. It is this overcrowding and hustling that gives the comic touch—this frantic haste of mobs of politicos, all on the run after being caught short, all spouting speeches and promises as they hurry in and out of the public view in a helter-skelter of cross-directions and cross-purposes. Invectives are being hurled, and even objects. While being televised at a country political meeting the other night, ex-Minister and present candidate François Mitterrand was hit on the nose by a pear, and bled freely before the camera. Already eleven of the national political parties have dropped from sight, as if they had fallen through a trapdoor. Even the names of the fantasy midget parties that only Frenchmen could father seem unusually odd this year. One outside Paris that has half a dozen candidates running is called the Witness of Christ Party, and is politically opposed to people's taking medicine. Another party calls itself the League of Consumers in Favor of Lowering the Cost of Living and Stabilizing the Franc. Still another, more vengeful, is named Social Solidarity

Against Former Deputies. The most optimistic little party is simply called the Fifth Republic.

It was well known in advance that Mendès-France would lead a new Left—his Republican Front—in a try for a majority to effect the vitally needed reforms in whatever sort of Parliament is hatched out in 1956. Nobody realized that a new extreme Right was also about to burst into the campaign—the Poujadistes. Pierre Poujade will be recalled as the husky, belligerent, thirty-five-year-old shopkeeper from St.-Céré, in southwestern France, who last winter roused a popular revolt among other little businessmen against paying what they call unfairly high taxes. Since then, his followers have cut their political eyeteeth pretty well all over France by getting into the Chambers of Commerce, which are elected bodies here. This month, exactly like Hitler when he was still a Fascist débutant, Poujade is shouting that his party is neither Left nor Right but only national. It is also anti-Semitic, anti-Communist, anti-democratic, anti-parliamentary, and pro-violence. A week or so ago, at a hysterical mass meeting, Poujade thundered, "When we come to power, if the deputies don't do their job right, we will hang them!" At Mendès-France's Salle de la Mutualité meeting for women voters, the well-organized Poujadiste strong-arm squad of hecklers yelled, "Mendès to the scaffold, Mauriac to the jackals [because of his democratic Moroccan sympathies], and Herriot to the Panthéon!," this last meaning that the octogenarian Radical Socialist leader is so old and dead that he should be entombed. At Poujade's Salle Wagram meeting, on mounting the platform he immediately did his regular striptease act, pulling off his windbreaker, scarf, and sweater and then rolling up his shirtsleeves, before firing his hearers with the declaration that "Our ancestors cut off the head of a king for doing far less than the men who govern us today!" As enthusiasm mounted, a woman devotee screamed, "Poujade is Jeanne d'Arc and Henri Quatre both!" In the tough style he affects, he finally called not only on his shopkeepers but on underpaid workmen and the ever dissatisfied peasantry to vote the Poujade ticket, "to save their guts and France." It is now expected that the Poujadistes will elect about ten deputies to the new Parliament—not bad for a beginning.

Registration has been the heaviest in the history of the two modern Republics, and the voting should be, too. Once more, it is Mendès-France who has galvanized the whole political scene. His speeches are dramatic and damning against the late government leaders. He is still attacking the curse of drink in France, and "the kingdom of alcohol" ruled by the industrial-alcohol trust. Though middle-class Left liberals are always saying that what France needs is a leader, some now say that they are afraid to vote for leader Mendès because he may be too autocratic. Others, being anti-Semitic, say that they are afraid to vote for him because voting for him, a Jew, will increase anti-Semitism here—a fine case of upside-down logic. The best anti-Communist political wisecrack of the campaign was made by Socialist chief Guy Mollet, who said disdainfully, "French Communists are not Left, they are East." Conservative Antoine Pinay, addressing a small, refined group of youths in a salon at the Hotel Lutétia, opened by remarking, "Someone has said that politics is the art of making possible that which is necessary." "Valéry—Paul Valéry, the poet—said it!" one listening youth shouted impatiently. Pinay, who was pinch-hitting for his more important and now ultraconservative running mate Edgar Faure, also told the young men, "Military service is out-of-date—just a few months of it will be enough." As further flattering planks in the Faure youth program, he promised them scholarships and organized part-time jobs, poverty among university students and the need to earn while studying being a terrible education problem here. He also promised government loans so that youth could get married while still young enough to enjoy marriage, with the loans being repaid in part, on a sort of installment plan, by the birth of each child. Oh, this is undoubtedly an extraordinarily entertaining French election campaign!

INDEX

Index

Mauriac, François (*continued*)
  clash with Cocteau, 168–169
  and colonials, 214
  Gide obituary, 136
  and *La Table Ronde,* 214–215
  as "Lighthouse," 117
  and Resistance, 17, 18
  and Sagan, 236
Maurras, Charles, 116, 184–185
Mayakavasky, Vladimir, 411–413
Mayer, René, 190–191, 194, 205,
  218, 220
McCarthy, Senator Joseph, 201, 202
"Mémoires de Guerre" (de Gaulle)
  Vol. I, 252–253
Mendès-France, Pierre
  background of, 229
  and campaign of 1955, 301–302
  and E.D.C., 244–245
  and German rearmament,
    258–260
  and Indo-China, 217
  and milk, 254, 263
  1954 program, 228
  overthrown, 264–265, 266
  and Premier Faure, 298, 299–300
  program of, 206, 208
  and Tunisia, 240–241, 272
  vote of confidence in, 247–249
Métro, cost of ride, 96
  and La Grande Nuit, 92
  and national strike, 211
  owner of, 56
  postwar, 15, 22, 47, 58
  strike of workers, 137
Milhaud, Darius, 219
Miró, Joan, 63
Mistinguett, 106
Moch, Jules, 108, 109
Mollet, Guy
  and German rearmament, 260
Monde, Le

on national strike (1953), 213
Mongibeaux, Paul, 126
  and Laval, 42–43, 44
  at Pétain trial, 32, 34, 38
Montand, Yves, 77, 113, 200
Moreno, Marguerite, 55
Mornet, Prosecutor, 32, 33, 40
Morocco
  Ben Moulay Arafa and, 214, 215
  and Grandval, 287
  massacre in, 291, 316
  and Matisse, 250–251
  Mohammed II, 170
"Mort de l'Homme, La" (Cocteau),
  61–62
Moysès, Louis, 101–102

NATO, and German rearmament,
  273–274
"Nausea" (Sartre), 49
"Neutralism," 134–135
Newspaper strike, 75, 76
North Africans, and Bastille Day
  (1955), 283–284
Nouvelles Littéraires, 116–117

Oïstrakh, David, 209
Opéra
  orchestra strike, 114
Opéra-Comique, 80, 114
Ormandy, Eugene, 274–275
Ortolans, 180–181

Paris
  two thousandth birthday, 147, 150
Paris, Comte de, 45
Paris *Herald Tribune,* 293
Paris-Match
  Anti-Americanism and, 184